Chemistry and Physics of Carbon

VOLUME 24

Chemistry and Physics of Carbon

A Series of Advances

Edited by

PETER A. THROWER

Department of Materials Science and Engineering
The Pennsylvania State University
University Park, Pennsylvania

VOLUME 24

CRC Press
Taylor & Francis Group
Boca Raton London New York

CRC Press is an imprint of the
Taylor & Francis Group, an **informa** business

CRC Press
Taylor & Francis Group
6000 Broken Sound Parkway NW, Suite 300
Boca Raton, FL 33487-2742

First issued in paperback 2019

© 1994 by Taylor & Francis Group, LLC
CRC Press is an imprint of Taylor & Francis Group, an Informa business

No claim to original U.S. Government works

ISBN-13: 978-0-8247-9091-2 (hbk)
ISBN-13: 978-0-367-40220-4 (pbk)

The Library of Congress Cataloged the
First Issue of This Title as Follows:

Chemistry and physics of carbon, v. 1-
    London, E. Arnold; New York, M. Dekker, 1965-

        v. illus. 24 cm

    Editor: v. 1- P. L. Walker

    1. Carbon.   I. Walker, Philip L., ed.

QD181.C1C44   546.681

Library of Congress   1   66-58302

**Visit the Taylor & Francis Web site at
http://www.taylorandfrancis.com**

**and the CRC Press Web site at
http://www.crcpress.com**

Preface

In spite of the discovery of diamond-like carbons and fullerenes, each with its own fascinating and potentially important properties, the production of bulk carbons and graphites remains the major industrial focus of carbon research. Much has been learned about the mechanisms of carbonization and graphitization since the publication of a landmark review by Brooks and Taylor in Volume 4 of this series in 1968, and it is appropriate that, 25 years later, we take a fresh look at the subject. Two chapters in this volume, one by Greinke and the other by Mochida et al., look at the carbonization process from different perspectives. I am sure readers will agree that, between them, these authors provide a useful account of our current understanding of a very complex process.

The editor is always eager to receive contributions on diamond, although only a few of the previous chapters in this series have focused on this important carbonaceous material. The fact that diamond can have a high thermal conductivity is often overlooked by those not directly involved with the material. The contribution by Morelli on the thermal conductivity of diamond is certainly timely as we hear more and more about potential uses for diamond thin films, and it should be of interest to all involved in the physical properties of carbon materials.

The final chapter in this volume deals with the ever-important topic of carbon surfaces, which figure not only in combustion and reactivity studies but also in adsorption, catalyst, and catalyst support applications. Leon y Leon and Radovic have compiled an extensive, valuable bibliography on this subject and have given a comprehensive review of current knowledge; there will undoubtedly be much of interest to all carbon scientists.

It hardly seems possible that I have now been involved with this series for twenty years. Sometimes it has been a real effort to assemble enough material for a volume, while on other occasions chapters have been submitted "out of the blue." I am always willing to entertain suggestions for and offers of new contributions. The continued success of this series must depend on its readers.

Peter A. Thrower

Contributors to Volume 24

Ken-ichi Fujimoto Technical Administration Division, Research and Development Laboratories, Nippon Steel Chemical Company, Ltd., Kitakyushu, Fukuoka, Japan

R. A. Greinke Parma Technical Center, UCAR Carbon Company, Inc., Parma, Ohio

Carlos A. Leon y Leon D.* Department of Materials Science and Engineering, The Pennsylvania State University, University Park, Pennsylvania

Isao Mochida Institute of Advanced Material Study, Kyushu University, Kasuga, Fukuoka, Japan

Donald T. Morelli Physics Department, General Motors Research and Development Center, Warren, Michigan

---

Current affiliation: Quantachrome Corporation, Syosset, New York

Takashi Oyama   Cokes Quality Control Group, Koa Oil Company, Ltd., Kuga-gun, Yamaguchi, Japan

Ljubisa R. Radovic   Department of Materials Science and Engineering, The Pennsylvania State University, University Park, Pennsylvania

Contents of Volume 24

Contents of Other Volumes

1

Early Stages of Petroleum Pitch Carbonization—Kinetics and Mechanisms

R. A. Greinke

UCAR Carbon Company, Inc., Parma, Ohio

I. INTRODUCTION

This review describes the early stages of carbonization of petroleum
pitches produced from decant oils. Decant oil-based petroleum pitches,
derived initially from the fluid catalytic cracking of petroleum residues,
are used in preparation of numerous useful carbon and graphite products
[1]. Petroleum pitch, for example, has been employed for many years as
an impregnation pitch to densify and strengthen the final graphite artifact
[2]. As a result of many years of analytical testing of petroleum pitches
in carbon research laboratories, considerable fundamental knowledge
of these materials was obtained. Excellent reviews of the advanced
analytical techniques used for characterizing pitches and carbonaceous
materials were made by Zander [3] and Lewis [4].

In 1965 Brooks and Taylor discovered the mesophase state as an
intermediate in the transformation of pitch to coke [5,30]. This fundamental discovery led to the development of a process for high-modulus
carbon fibers, attributed much to the pioneering work of L. S. Singer [6].
During the 1970s and early 1980s, many workers at the Carbon Products
Division of the Union Carbide Corporation (the former name of UCAR
Carbon Company, Inc.) were involved in producing significant quantities
of continuous filament high-modulus carbon fibers from spinning
mesophase petroleum pitch [7]. Although a plethora of methods existed
in our laboratory for the characterization of petroleum pitches, an
in-depth understanding of the transformation of isotropic petroleum
pitch to mesophase pitch and subsequently semicoke presented new
challenges.

Before proceeding to the detailed discussions of early stages of
petroleum pitch carbonization, some basic definitions will be reviewed.
A description of the fluid catalytic cracking (FCC) process for the

conversion of petroleum fractions into high-value gasoline and light fuel oil components is described by Franck and Stadelhofer [8]. The highest boiling fraction of the overhead products is the precursor of decant oil. This high boiling fraction contains entrained catalyst particles that can be removed. Decant oil is the liquid material that has been separated from the catalyst particles. The decant oil can be converted to a petroleum pitch by distillation. The term *petroleum pitch* used in this review therefore refers to the residuum carbonaceous material obtained from the catalytic cracking of petroleum distillates or residues [9,10]. The term *anisotropic pitch* or *mesophase pitch* denotes pitch composed of molecules with an aromatic structure that through physical interaction have associated together to form nematic liquid crystals. The term *isotropic pitch* signifies pitch composed of molecules that are not associated into liquid crystals. The term *mesogens* means mesophase-forming materials or mesophase precursors [9,10]. A *mesogenic pitch* has not yet transformed into mesophase. *Mesogenic molecules* are found in a mesogenic pitch.

Petroleum pitches derived from a decant oil are complex in constitution and are composed of mixtures of hundreds of compounds as observed by mass spectroscopy [11]. Polycyclic aromatic hydrocarbons (PAH) comprise the dominant class of compounds present in a petroleum pitch [12]. Other classes are alkylated PAH, partially hydrogenated PAH, and heterocyclic aromatic hydrocarbons [12]. The development of the mesophase stage from this complex chemical mixture during the thermal conversion of a petroleum pitch is shown in Fig. 1 [82]. The figure contains polarized light photomicrographs of a decant oil petroleum pitch as it is heat treated at 400°C. In a very general description of the conversion process, the heat treatment results in volatilization of lower molecular weight components (inhibitors of mesophase formation or nonmesogens) and polymerization of the more reactive species. As a result of chemical polymerization reactions increasing the average molecular weight of the pitch, structural order (a physical transformation process) is detected by polarized light microscopy [13]. The carbonaceous mesophase develops initially as small anisotropic spherules that precipitate from the original isotropic pitch phase. As the heat treatment progresses, the spherules grow and then coalesce to form large bulk anisotropic regions that separate from the lower-density isotropic pitch phase and slowly settle. Subsequently, as the chemical reactions continue, the entire pitch is transformed to a fusible mesophase state.

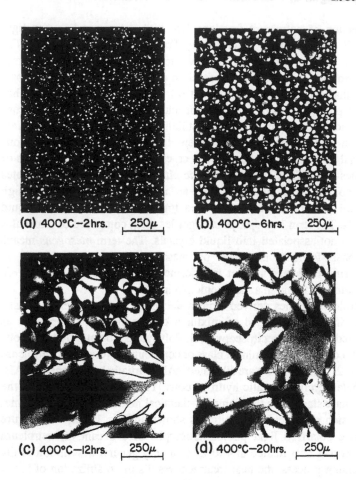

(a) 400°C−2hrs.   250μ   (b) 400°C−6hrs.   250μ

(c) 400°C−l2hrs.   250μ   (d) 400°C−20hrs.   250μ

FIG. 1   Development of mesophase during 400°C heat treatment of petroleum pitch as observed by polarized light microscopy of polished sections at room temperature: (a) 2 h, (b) 6 h, (c) 12 h, (d) 20 h. (Reprinted with permission from Ref. 82.)

Additional heat treatment results in additional polymerization reactions of the mesophase molecules and leads to the formation of a very high softening point semicoke and then an infusible anisotropic coke. The unique characteristic of mesophase formed from decant oil-based pitch is its low viscosity in the molten stage. The low viscosity of the mesophase allows spinning of this material into high-modulus fibers at a relatively low temperature [14].

Although the conversion of isotropic pitch to mesophase pitch could be observed by microscopy, many fundamental questions were asked in our laboratory regarding the nature of this transformation. What size molecules react to form the mesophase? What are the rate constants of the reactive molecules in the pitch as a function of molecular size? Do the rate constants for the polymerization reaction change during the transformation from isotropic pitch to the fluid mesophase? What is the molecular constitution of the mesophase? What are the constitutions of the coexisting mesophase and isotropic phase during the transformation? What is the mechanism for mesophase sphere growth? To answer these and numerous other questions, new analytical methodology had to be developed and applied to conventional characterization techniques.

II. ANALYTICAL METHODOLOGY

Key analytical techniques used in our studies were gel permeation chromatography (GPC), high-temperature centrifuging, and proton nuclear magnetic resonance spectroscopy (^{1}H-NMR). Although the conventional analytical techniques of GPC and ^{1}N-NMR had previously been applied to evaluate pitches, special procedures had to be developed because of the insolubility of mesophase pitch in conventional GPC and ^{1}H-NMR solvents. At the time of this writing, the combination of GPC and high-temperature centrifuging techniques used to obtain molecular weight distributions of the coexisting isotropic and mesophase fractions has been accomplished only by the Parma laboratory and has provided key and unique insights to the understanding of polymerization reactions and mesophase formation mechanisms occurring during the early stages of pitch carbonization. A brief description of these special methodologies and their historical development follow.

A. Gel Permeation Chromatography

Edstrom and Petro [15] first used GPC in our laboratory to measure the molecular size distribution of pitches using tetrahydrofuran as an eluent. Lewis and Petro [16] reported the first use of GPC to obtain molecular weight distributions of polymerized pitches using toluene as an eluent. However, nonideal elution behavior for the pericondensed polynuclear aromatic molecules, known to be present in petroleum pitches [12,17,18], has been observed in tetrahydrofuran [15], toluene [16,19], and other commonly used GPC eluents such as benzene [20] and methylene chloride [21]. The use of these common GPC eluents results in a chromatographic method that does not elute molecules by molecular size. The elution of pitch molecules by size is a condition essential for accurately answering many of the foregoing questions regarding the early stages of petroleum pitch carbonization. Fortunately, several eluents including quinoline [22] and 1,2,4-trichlorobenzene [19], where near-ideal molecular size elution behaviors exists, were found.

The next major problem confronting the chromatographic characterization of mesophase pitch was the insolubility of high molecular weight polymerized molecules even in the best eluent, quinoline. This barrier was overcome by applying reductive hydrogenation [23,24] and reductive ethylation [25] techniques that had previously been used to solubilize coal. The reduction techniques introduce hydroaromatic rings and short side chains into the structure without any significant altering of the effective molecular size. (See Fig. 2 for the chemical changes occurring in a hypothetical petroleum pitch molecule after reductive hydrogenation and reductive ethylation.) The reduction techniques are known to fracture carbon–oxygen bonds (ethers) in coals. Extensive use of reduction techniques with mesophase pitch in our laboratory has shown that carbon–carbon bonds in large pitch molecules do not fracture. Hence, depolymerization of the large pitch molecules has never been detected.

The detailed experimental solubilization techniques for subsequent chromatographic evaluation in quinoline have been described by Greinke [26]. The detailed experimental procedures for the chromatographic analysis of mesophase pitch using trichlorobenzene eluent have been reported by Greinke and Singer [27]. A GPC calibration curve for quantifying the molecular weight distribution of polymeric petroleum pitch (Fig. 3) was obtained by Greinke and O'Connor [28] by fractionating the reduced polymeric pitch and subjecting the collected fractions to vapor phase osmometry for accurate number average molecular weight. The most remarkable feature of the calibration curve of Fig. 3 for petroleum pitch is that the exclusion limit of the chromatographic columns approached a molecular weight of only 2000, a relatively low molecular weight. Compared to other polymer systems

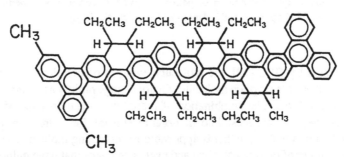

CH₃

HYPOTHETICAL PITCH MOLECULE

CH₃

REDUCTIVELY HYDROGENATED MOLECULE

REDUCTIVELY ETHYLATED MOLECULE

FIG. 2 Hypothetical structures of a high molecular weight petroleum pitch molecule before reduction, after hydrogenation, and after reductive ethylation. (Structures suggested by Dr. E. M. Dickinson.)

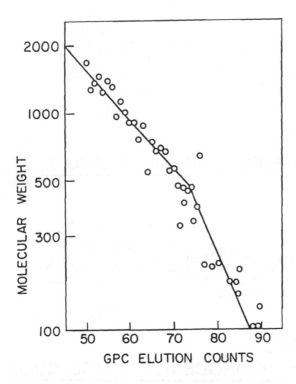

FIG. 3 Gel permeation chromatographic calibration curve for quantifying the molecular weight distribution of petroleum pitch. (Reprinted with permission from Ref. 28. Copyright 1980, American Chemical Society.)

separated by GPC [29], the molecular weights of petroleum pitch and reduced petroleum pitch molecules do not change rapidly as a function of elution volume. The combination of reductive ethylation coupled with quinoline, the strongest known organic solvent for petroleum pitch, resulted in a technique that is applicable for obtaining molecular weight distribution of mesophase pitch and even semicokes that have quinoline-insoluble contents approaching 100% before reduction. The possibilities for studying the mechanisms and chemical kinetics of the underlying reactions that lead to the physical transformation of an isotropic pitch to anisotropic pitch as observed by polarized light microscopy became very apparent.

B. High-Temperature Centrifugation

The carbonaceous mesophase is slightly more dense than the isotropic phase from which it forms and therefore slowly settles to the bottom of a nonagitated reaction vessel at high temperatures [24]. The enhancement of the rate of phase separation using high-temperature centrifugation was therefore obvious and was developed by several investigators. Rand and co-workers [31,32] used high-temperature centrifugation to study the glass transition of the separated phases, whereas Chen and Diefendorf [33] used high-temperature centrifugation to evaluate solubilities of the separated fractions. In our laboratory, Singer et al. [34] developed a high-temperature centrifuge capable of operating up to 500°C at a centrifugal acceleration of 600 g. These authors indicate that the high-temperature centrifugation of mesophase pitches can provide information not readily obtainable by other techniques. The characterization of the separated phases in mesophase pitches will lead to a better understanding of molecular, structural, and distributional requirements of mesophase formation.

Figure 4 shows a polarized light photomicrograph of the isotropic–anisotropic phase interface in a petroleum mesophase pitch after separation by high-temperature centrifugation. Although some interphase contamination is apparent in the photomicrograph (i.e., mesophase trapped in the isotropic and isotropic trapped in the mesophase), the amounts were relatively small and lead only to small errors [27]. The description for obtaining the separated phases for constitutional analysis are reported in detail [34].

C. Proton Nuclear Magnetic Resonance

Although ^{1}H-NMR has been used extensively for the chemical characterization of petroleum pitches and tars [18,35,36], a polymerized pitch that contains mesophase cannot be analyzed by conventional solution ^{1}H-NMR, because of limited solubility in standard solvents, such as CS_2 and deuterated pyridine. Generally, a mesophase pitch derived from a petroleum pitch contains about 55% pyridine insolubles. A very simple ^{1}H-NMR technique was employed for the complete characterization of polymerized petroleum pitches with pyridine-insoluble contents as high as 80% [37]. The polymeric pitch is dissolved in a mixed inorganic solvent, consisting of sulfur monochloride (S_2Cl_2) and sulfuryl chloride (SO_2Cl_2).

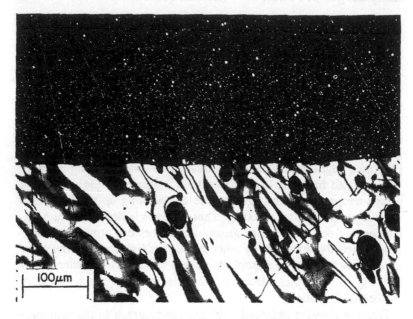

FIG. 4   Polarized light photomicrograph of the centrifuged separated phases of a mesophase pitch. (Reprinted with permission from Ref. 68, by Gordon and Breach Science Publishers.)

The proposed solubilization mechanism of the polymeric pitch involves a two-step reaction. The first step is the oxidation of pitch molecules to form radical cations, and the second step is the nucleophilic chlorination of the radical cations [37]. Although solubilization occurs via the addition of chlorine to the polynuclear aromatic rings, structural information of polymerized pitches can still be obtained. The main uses of this technique are to determine the ratio of aromatic to aliphatic protons in polymeric pitch and to monitor the kinetics of the pitch polymerization reaction of the methyl protons. Our experience with the technique indicates that it can be applied quantitatively only to polymeric pitches or to pitches that do not contain any pitch volatiles (molecular weight less than 400). The technique cannot be applied to low molecular weight model compounds or to tars. Other instrumental NMR techniques, namely high-temperature or fused-state NMR [38–42] and solid-stage NMR [43], have been reported for structural characterization of polymeric pitch.

III. PITCH POLYMERIZATION KINETICS DURING MESOPHASE FORMATION

A. General Description

The changes in the molecular weight distribution (MWD) of the petroleum pitch after heat treatment at 400°C for reaction times between 0 and 16 h have been reported [26] and are shown in Fig. 5. The pitch, after heat treatment for 16 h, contains 100% mesophase and 60% pyridine insolubles. Since the polymerization was accomplished at atmospheric pressure with a sparging gas, the molecules with molecular weight (MW) less than 400 are distilled rapidly as observed in Fig. 5. After distillation, the petroleum pitch contains negligible amounts of mesophase. Between 2 and 16 h, the mesophase content increases to 100%, whereas the number average MW of the heated pitch increases from 721 to 880, an increase that is almost entirely attributed to polymerization since most of the low MW volatiles were distilled prior to 2 h. The MWD curve possesses a very large peak centered at about MW 580. This peak, characteristic of all petroleum pitches evaluated by the author using GPC, also was observed at MW of 580 by Mochida using field desorption mass spectroscopy [44]. After 16 h of heat treatment, a new peak, although less sharp, is observed to form at MW 1200. The 1200 MW peak would be expected in a polymerization process since it is approximately double the MW of 580. As a result, one can broadly classify the molecules near the two peaks as "monomer" and "dimer" to describe simplistically the polymerization process occurring in a petroleum pitch during mesophase formation.

The MWD data of Fig. 5 show that many of the original size pitch monomers (MW of 400–800) remain in the mesophase pitch and that about only half of the monomers have reacted during the isotropic-to-mesophase transformation after 16 h of heat treatment. The molecular weight distribution of the mesophase pitch (16-h sample), as shown in Fig. 5, is broad and ranges between 400 and 2000. Similarly, the MW range of a decant oil-based mesophase pitch measured by Mochida et al. [45,46] using vapor phase osmometry varied between 400 and 3000. Similarly, the molecular weight range of mesophase pitch obtained from Ashland 240 pitch (a decant oil-based pitch) by Chen and Diefendorf varied from 230 to greater than 1400 [33].

An interesting observation of the MWD shown in Fig. 5 is that significant amounts of higher molecular weight pitch (approximately 30%

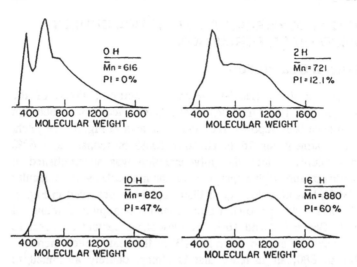

FIG. 5 Molecular weight distributions obtained from GPC curves of a petroleum pitch as a function of heat-treatment time at 400°C. (Reprinted with permission from Ref. 26, Copyright 1986, Pergamon Press Ltd.)

by weight) greater than 1000 are present in the 2-h reacted sample, but molecules with molecular weights near 2000 are not present even after 16 h. This visual observation suggests that reactions of the 1000 MW and larger molecules with each other are negligible during the transformation to mesophase pitch. The reactivity of the petroleum pitch molecules as a function of molecular size will be presented in the next section.

Although the underlying chemical reactions of petroleum pitch that lead to the physical mesophase formation have been described as a product of consecutive reactions of pitch molecules [47], the GPC data of Fig. 5 indicate a single polymerization step of pitch monomers to dimers. According to Fig. 5 (16-h sample), the mesophase obtained from a decant oil-based petroleum pitch is composed of a mixture of pitch monomers and pitch dimers with the most intense weight near molecular weights of 600 and 1200. The simple reaction scheme occurring at 400°C at atmospheric pressure based on the GPC data can be described as follows:

$$\{A + a\} \xrightarrow{\ -(a)\ } \{A\} \xrightarrow{\ k_1\ } \{PA\}$$

$$\{MP\} = \{A\} + \{PA\}$$

where

{A} are pitch monomers with a molecular weight range of between 400 and 1000 and a peak MW of 580.

{a} are pitch volatiles (nonmesogens or mesophase inhibitors) with a molecular weight less than approximately 400.

k_1 are the rate constants of the reactive molecules in pitch at 400°C.

{PA} are pitch dimers with a molecular weight range of between 1000 and 2000 and a peak molecular weight of approximately 1200.

{MP} is mesophase pitch, which contains pitch monomers present in the starting pitch and dimers, many of which are formed by the thermal treatment.

The type of chemical bond formed in the polymerized pitch molecules (the 1200 MW dimers of Fig. 5) has been the subject of several investigations. Delhaes et al. [48], using diamagnetic susceptibility measurements, have shown that the polymerized molecules in mesophase pitch are composed of not too highly condensed pitch monomers connected by single chemical bond. Mochida [46], after applying the depolymerization reaction of phenol and p-toluene-sulfonic acid to mesophase spheres, concluded that the nature of the chemical bonds in the polymerized or dimerized molecules were predominately aryl–aryl bridges with smaller amounts of methylene bridges. Recent quantitative mechanistic studies of the thermal cracking and gas formation reactions of alkyl groups on aromatic molecules (greater than 400 MW) present in decant oil-based mesophase pitch indicate that the side chains fracture at the alpha position to the aromatic ring, which leads to the formation of aryl–aryl bridges in the pitch dimers and not methylene bridges [49].

B. Rate Constants

What is the reactivity of petroleum pitch molecules as a function of molecular size? With the development of a quantitative GPC technique for determining the molecular weight distribution of mesophase pitch [26], this question can now be answered. Before describing the results of Greinke [26], a brief summary of some interesting prior kinetic studies will be mentioned.

Marsh and Walker [50] indicated in a previous review that kinetic studies of pitch polymerization are limited and that interpretations of kinetic data are difficult since the molecular components of pitch react at different rates. Perhaps the simplest and most common technique used for studying the kinetics of pitch polymerization is the measurement of changes in pitch solubility. Lewis presented a review paper on the chemistry of carbonization and showed the importance of dehydrogenative polymerization on the carbonization process [4]. The dehydrogenative polymerization reactions decrease the pitch solubility in solvents such as pyridine and toluene.

A number of studies in which the kinetics are monitored by measuring changes in solubility have been reported [51,52,53]. The polymerization reactions follow first-order kinetics based on solubility measurements [53]. An electron spin resonance study of the kinetics of carbonization by Singer and Lewis [53] show that the dehydrogenative condensation reactions are the carbonization reactions, which lead to increasing concentrations of stable free radicals. The free-radical kinetics for the transformation of pitch show a first-order dependence similar to the formation of pitch insolubles [53].

Only two prior studies that attempted to define the reactivity of petroleum pitch as a function of molecular size were found. After determining gradual decreases in the average MW of the chloroform-soluble fraction extracted from a series of heat-treated pitches, Brooks and Taylor [54] concluded that their results "seem to indicate, in the early stages of carbonization (during mesophase formation), a preferred condensation of compounds of higher MW." In a similar experiment, Lewis and Didchenko [55] concluded that the highest molecular weight components of the toluene-soluble fraction extracted from a series of heat-treated pitches were polymerizing most rapidly. The GPC studies [26], however, conflict with these results, namely, that the rate constants of the larger molecules in a petroleum pitch decrease with increasing molecular size.

A summary of the GPC polymerization kinetics for one petroleum pitch follows. However, a number of petroleum pitches and coal tar pitches, which result in low-viscosity fluid mesophase, were found by the author to have similar GPC polymerization patterns.

1. Rate Constants of the 400–700 MW Molecules

By using the sequence of molecular weight distributions such as those shown in Fig. 5, one can calculate the first-order rate constants for pitch polymerization at 400°C as a function of molecular size [26]. The study showed that the rate constants of the molecules with MW of 400–700 were essentially similar. For example, an average first-order rate constant at 400°C of 0.73×10^{-5} s^{-1} was calculated for the 400–560 MW range, whereas an average rate constant at 400°C of 0.68×10^{-5} s^{-1} was calculated for the 560–700 MW range. Combining the results, one measures the average kinetics of the reaction of the 400–700 MW range of molecules (Fig. 6a). A straight line plot was obtained, which suggests that the polymerization reaction obeys first-order kinetics with an average rate constant of 0.70×10^{-5} s^{-1} at 400°C. The first-order kinetics observed using GPC analysis is consistent with the results of other research. The first-order reaction was previously interpreted that the rate-determining step is the formation of a free radical on a single molecule [50,53]. Yamada et al. [56] further suggested that the rate-determining step is radical formation because of the similarity in values

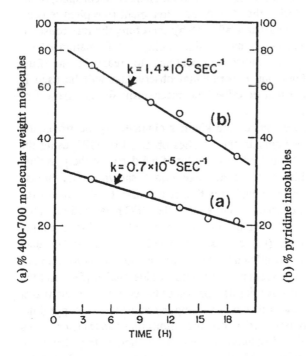

FIG. 6 Kinetics of the reaction of (a) the 400–700 MW molecules and (b) the pyridine-soluble fraction of a petroleum pitch as a function of time at 400°C. (Reprinted with permission from Ref. 26, Copyright 1986, Pergamon Press Ltd.)

of rate constants and activation energies for radical formation by fission of C–C bonds between aromatic and aliphatic carbon atoms. Lewis and Singer [57] more specifically indicate in their review article that the first step in the carbonization sequence is formation of a sigma free radical that has low stability, that cannot be detected by electron spin resonance, and will react immediately to form higher molecular weight polyaryl polymers. The data of Fig. 5 suggest that the reactive sigma free radicals are formed in the monomer molecules centered at MW of 580 to form the dimer molecules, centered at MW of 1200.

Do the rate constants for the polymerization reaction change during the transformation from isotropic pitch to the fluid mesophase? Or, asked in another manner, does the molecular order or the physical alignment of the molecules that takes place during polymerization influence the rate constants of the monomer molecules in the pitch? The answer is no. The data of Fig. 6a show that the slope (rate constant) of the plot is unchanged during the 19 h of heat treatment (the pitch is completely converted to mesophase after 16 h); therefore, the rate constants of these size molecules do not change during the formation of the mesophase. These size molecules continue to react in the fluid mesophase. This important experimental observation will be used in a subsequent section that describes the mechanism for the growth of mesophase spheres.

Greinke [26] also compared the kinetics obtained by the frequently used solubility measurements and kinetics obtained by GPC using the same series of heat-treated pitches. A plot of the decay of the pyridine-soluble fraction (obtained by soxhlet extraction), shown in Fig. 6b also suggests a first-order reaction, but with a rate constant twice that of the GPC measured rate constant. Compared to the GPC measured rate constants, the doubled value of the rate constant obtained for formation of pyridine insolubles (PI) can be attributed to factors other than polymerization that also influence the PI content, such as conversion of hydroaromatic rings to aromatic ring, loss of side chains [58], and the "cosolvent effect" [59]. A brief description of the cosolvent effect is that pitch molecules aid each other in dissolving in a solvent such as pyridine. During the course of polymerization, pitch molecules can transfer from a soluble fraction to an insoluble fraction even though they have not reacted, as a result of reduced pitch cosolvents in the matrix. The influence and significance of this cosolvent effect will be examined in subsequent discussions.

2. *Rate Constants of the 700–1200 MW Molecules*

The results of Fig. 6a, which quantify the rate constants of the 400–700 MW molecules, coupled with the results observed in Fig. 5, which suggest that molecules larger than 1000 MW react negligibly during mesophase formation, indicate that there is a rapidly decreasing polymerization reactivity for molecules between 700 and 1100 MW. In order to determine the rate constants of pitch molecules in this MW range, Greinke first extracted the petroleum pitch [26]. The extraction procedure removed many of the 400–700 MW molecules, which allowed one to study the reaction of the larger 700–1100 MW molecules, with minimum interference from the reaction of the 400–700 MW molecules. The MW distributions, obtained for the series of samples obtained by heat treating the extracted pitch at 400°C, are shown in Fig. 7. Again, the failure to produce molecules with MW of 2000 and larger in this series of heat treatments indicates a unique stability during mesophase formation of the larger dimer molecules present in the petroleum pitch. The first-order kinetic rate constants, calculated from the curves of Fig. 7, are

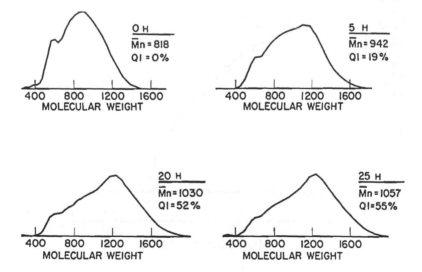

FIG. 7 Molecular weight distribution obtained from GPC curves of an extracted petroleum pitch at different reaction times at 400°C. (Reprinted with permission from Ref. 26, Copyright 1986, Pergamon Press Ltd.)

shown in Fig. 8a. The rate constants decrease significantly beginning at MW 800 and become almost negligible at MW 1100 [26]. The stability of the larger molecules occurs in molecules containing stable pi free radicals [57]. The concentrations of stable free radicals in petroleum feedstocks were shown to increase with molecular weight [60].

As mentioned previously, these results differ from other workers who reported increasing rate constants of the larger molecules [54,55]. The discrepancy can be explained by the cosolvent effect in which the larger molecules initially present in the solvent-soluble fraction at early heat-treatment times transferred to the solvent-insoluble fraction at longer heat-treatment times even though they did not polymerize. This transfer,

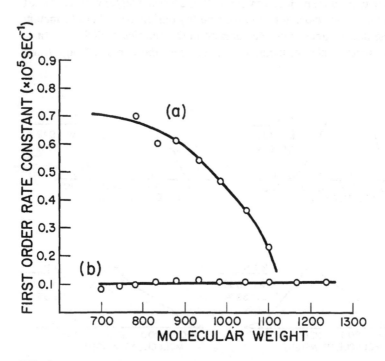

FIG. 8 First-order rate constants of different size molecules in an extracted petroleum pitch at 400°C: (a) 5–25 h, 19–55% quinoline insolubles and (b) 25–161 h, 55–99% quinoline insolubles. (Reprinted with permission from Ref. 26, Copyright 1986, Pergamon Press Ltd.)

owing to diminished amounts of pitch cosolvents, resulted in the illusion of faster reactivity of the larger molecules in the solvent-soluble fraction.

Combining the results of Figs. 6a and 8a, one observes that the dehydrogenative polymerization reactions of a petroleum pitch (which lead to the transformation to mesophase pitch) involves the 400–1100 MW molecules. This reactivity "window" was observed by the author in other petroleum pitches, coal tar pitches, pyrolysis tar pitches, etc. Only the magnitude of the rate constants as a function of molecular weight, the breadth of the window, and the MW where the pitch reactivity diminishes to negligible values varied for each pitch source. These differences are related to the chemical constitution of the various pitches.

As another example, the first-order rate constants are shown in Fig. 9 as a function of molecular weight for the polymerization of the molecules present in pitches produced from the model compounds, naphthalene and

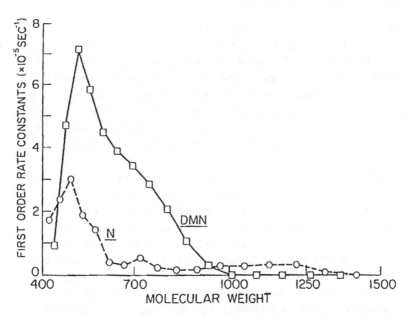

FIG. 9 First-order rate constants for the polymerization of different size molecules in a naphthalene (N) pitch and dimethylnaphthalene (DMN) pitch during mesophase formation at 420°C. (Reprinted with permission from Ref. 26, Copyright 1986, Pergamon Press Ltd.)

dimethylnaphthalene [61]. The data show the activating effect of the methyl substituent groups on both the magnitude of the rate constants and the molecular weight breadth of the reactive molecules in the window. The smaller molecules present in both model-compound-devolatilized pitches exhibit the greatest reactivity.

The stability of the larger molecules (dimers) present in pitches during mesophase formation is indeed fortuitous, since it ensures that the mesophase remains highly fluid over a wide range of reaction conditions and that coke particles do not form and precipitate in the fluid mesophase. The prolonged fluidity is important for many industrial processes using mesophase pitch. As will be shown in a subsequent section, the polymerization reactions of these larger molecules occur after the mesophase has been thermally advanced to a semicoke with a very high softening point.

C. Activation Energies

The activation energies for petroleum pitch polymerization have been measured by solvent extraction techniques [51,52,53]. The rate constants of pyridine-insoluble formation for a petroleum pitch were measured at five different temperatures by Singer and Lewis [53]. An Arrhenius plot of ln k (rate constant) versus $1/T$ (absolute temperature, K) for the formation of pyridine insolubles in the petroleum pitch (Fig. 10) resulted in an activation energy near 50 kcal/mol. This value of activation energy for petroleum pitch is typical of those obtained by workers using solvent extraction techniques.

Activation energies have also been measured by GPC for petroleum pitch molecules as a function of molecular weight [26]. The first-order rate constants calculated at 380, 400, and 420°C as a function of MW are given in Table 1, together with the corresponding activation energies. The activation energies as a function of molecular size were all approximately 50 kcal/mol, values similar to those obtained by solvent extraction techniques. The data of Table 1 suggest that it would be difficult to change the molecular weight distribution of a mesophase pitch significantly by changing the temperature of polymerization.

The data of Fig. 6a, which plots the rate constant of the 400–700 MW molecules, suggest that the pitch volatiles, particularly those with MW approaching 400, may have appreciable reactivity if

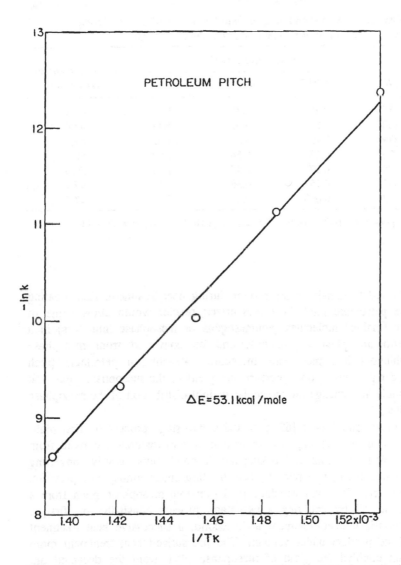

FIG. 10 Arrhenius plot of ln k versus $1/T$ for pyridine-insoluble formation in a petroleum pitch. (Reprinted with permission from Ref. 53, Copyright 1978, Pergamon Press Ltd.)

TABLE 1   Activation Energy of Petroleum Pitch Molecules during
Mesophase Formation as a Function of Molecular Weight

| MW | Rate constant ($\times 10^5$ s^{-1}) | | | Activation energy (kcal/mol) |
	380°C	400°C	420°C	
785	0.21	0.70	1.74	47.1
830	0.20	0.60	1.74	48.4
879	0.19	0.61	1.69	49.1
930	0.18	0.54	1.69	50.9
984	0.17	0.45	1.56	50.3
1042	0.15	0.36	1.18	46.4
1102	0.098	0.23	0.82	47.5

allowed to remain in the reactor during heat treatment. Heat treating the petroleum pitch in a pressurized reactor would allow some of the smallest molecules (nonmesogens or mesophase inhibitors) in a petroleum pitch to condense and be converted from mesophase inhibitors into mesogenic molecules. Reacting a petroleum pitch under pressure should increase the yield of the mesophase pitch and should also change the molecular weight distribution of the mesophase pitch.

Lewis and Moore [62] patented a two-stage process by first high-pressure heat treating low molecular weight distillates, derived from petroleum refining and coking and, second, subsequently converting it into a mesophase pitch by heat treating under atmospheric pressure. Similarly, Park and Mochida [63] prepared mesophase pitch from a decant oil by applying a two-stage process consisting of, first, a pressurized heat treatment and, second, a successive heat treatment of the product under vacuum. The pressurized heat treatment more than doubled the yield of mesophase pitch from the decanted oil. Lewis and Lewis [64] heat-treated 130°C softening point petroleum pitches under atmospheric pressure without sparging to a predetermined mesophase containing pitch and thereafter continued the heat treatment with sparging. Increased yields as high as 30% were reported.

IV. CONSTITUTION OF COEXISTING PHASES IN MESOPHASE PITCH DURING HEAT TREATMENT

What is the changing relationship between the coexisting phases of a petroleum pitch during heat treatment? What is the mechanism for the growth of the mesophase spheres as observed by polarized light microscopy? What is the mechanism of mesophase development during transformation? These questions will be answered in this section by focusing on the polymerization process of a petroleum pitch during mesophase formation. As mentioned previously, considerable insights pertaining to the understanding of the early stages of petroleum pitch carbonization were obtained by combining the techniques of high-temperature centrifuging for separating the coexisting phases with analytical methods to characterize the separated phases. Because of the many procedural steps of this approach, only a few studies of this kind have been accomplished with petroleum mesophase pitch [27,32,33].

Experimentally, representative samples are removed periodically from the stirred reactor during the transformation of isotropic pitch to mesophase pitch. Then the coexisting phases are separated via high-temperature centrifuging. Finally, the separated phases are characterized by using various analytical techniques [27].

A. Characterization by Solubility

Greinke and Singer [27] removed three samples of polymerized petroleum pitch from a stirred reactor at atmospheric pressure after 8, 11, and 15 h of heat treatment at 400°C. These samples contained 26, 44, and 64% mesophase. The weight percentages of pyridine insolubles in the three original samples and in the centrifuged separated phases are given in Table 2 [27]. The PI contents of the three separated mesophase fractions remained constant (approximately 56%) with reaction time, whereas the PI contents of the isotropic fractions increased significantly with time. The solubility data in Table 2 illustrate the unique phase partitioning occurring. For example, the isotropic phase from sample C contains 32.9% PI and essentially no mesophase, whereas the whole sample B contains about the same PI content (36.5%) but has 44.6% mesophase. These data indicate that it is impossible to obtain the same molecular weight constitution of the isotropic fraction by solvent frac-tionation of the mesophase pitch as it is by phase partitioning the mesophase pitch using centrifugation. These data are consistent with the

TABLE 2 Amount of Pyridine Insolubles and Short Side Chain Protons in the Whole Petroleum Mesophase Pitch and in the Separated Isotropic and Mesophase Fractions

Sample ID	PI (%)	Short side chain protons (%)
A. 8 h (26.3% mesophase)		
1. Whole	25.5	27.5
2. Mesophase fraction	55.6	25.8
3. Isotropic fraction	16.4	27.4
B. 11 h (44.6% mesophase)		
1. Whole	36.5	25.9
2. Mesophase fraction	55.9	23.9
3. Isotropic fraction	29.0	27.2
C. 15 h (64.3% mesophase)		
1. Whole	53.1	24.0
2. Mesophase fraction	55.9	23.4
3. Isotropic fraction	32.9	24.7

Reprinted with permission from Ref. 27, Copyright 1988, Pergamon Press Ltd.

work of Chwastiak and Lewis [65] who reported that the mesophase content of petroleum pitches cannot be equated to the amount of PI present in the petroleum pitch.

Similarly, Chen and Diefendorf [33], using centrifuging techniques with a heat-treated Ashland 240 pitch, reported that both the resulting coexisting isotropic phase and mesophase contain significant amounts of both toluene-soluble and toluene-insoluble fractions. The coexisting isotropic phase contained 41 wt% toluene-insoluble fraction that can form mesophase, whereas the mesophase fraction contained 44 wt% of a material that formed an isotropic melt. Chen and Diefendorf [33] used solubility theory to explain the qualitative behavior of the system.

B. Characterization by Size Exclusion Chromatography

The number average molecular weights, M_n, and the dispersions, $D_{z/n}$ (z average molecular weight divided by the number average molecular weight), for the three centrifuged, separated samples of Table 2 were reported by Greinke and Singer [27] and are given in Table 3. A plot of

TABLE 3 Number Average Molecular Weight, M_n, and Dispersions, $D_{z/n}$, of Whole Petroleum Pitches and Separated Phases

Sample	A (26.3% mesophase)	B (44.6% mesophase)	C (64.3% mesophase)
Whole pitch			
1. $\overline{M}_n$	797	852	863
2. $D_{z/n}$	1.28	1.28	1.32
Isotropic fraction			
1. $\overline{M}_n$	772	799	838
2. $D_{z/n}$	1.26	1.26	1.28
Mesophase fraction			
1. $\overline{M}_n$	902	926	903
2. $D_{z/n}$	1.32	1.30	1.32

Reprinted with permission from Ref. 27, Copyright 1988, Pergamon Press Ltd.

the M_n values of the separated isotropic and mesophase fractions as a function of total mesophase content in the original mesophase pitches is shown in Fig. 11. The results indicate that the MWDs of the separated mesophase fractions are invariant during the transformation and that the M_n of the separated isotropic fractions increase during the transformation. The M_n of the isotropic phase approaches that of the mesophase (greater than 900 amu) as the amount of mesophase in the pitch approaches 100% (Fig. 11). Similarly, by extrapolation to 0% mesophase, one calculates that the mesophase formation initiates when the isotropic phase has a number average molecular weight of approximately 725 amu. Using hot stage microscopy, Lewis [66] observed that the viscosity of the isotropic phase of a petroleum pitch was always less than that of the coexisting mesophase, an observation consistent with the data of Fig. 11 and Tables 2 and 3. In addition, Lewis et al. reported a nearly constant mesophase domain size by optical microscopy during the transformation [67], a result also suggesting that the molecular constitution of the mesophase is constant during the transformation.

The molecular weight distributions of the whole (original) mesophase pitch, the separated mesophase, and the separated isotropic phase from sample C (the 15-h heat-treated sample in Table 2 that contains 64.3% mesophase) were reported by Greinke and Singer [27] and are displayed

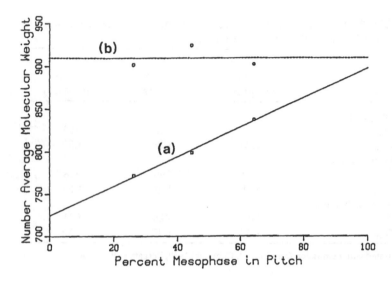

FIG. 11 Number average molecular weight of the coexisting isotropic (curve a) and mesophase (curve b) fractions in a petroleum pitch during the transformation to mesophase. (Reprinted with permission from Ref. 27, Copyright 1988, Pergamon Press Ltd.)

in Fig. 12. This figure illustrates that both the coexisting mesophase and isotropic phase contain similar size molecules ranging from 400 to 2000 amu. Greinke and Singer [27] also measured the partitioning of the molecules between the coexisting phases from the size exclusion chromatography curves that were normalized to a constant injection weight. The partition ratio was defined as the height (which is proportional to weight) of the size exclusion chromatography curve of the isotropic phase divided by the height of the SEC curve of mesophase at a particular molecular weight [22]. Hence, the partition ratios, which quantify the constitutional differences between the coexisting phases for samples A and C of Table 2, are plotted as a function of molecular weight in Fig. 13.

Both Figs. 12 and 13 illustrate that the corresponding anisotropic and isotropic phases of a well-sparged mesophase pitch contain similar size molecules and that only the amount of each size molecules varies in the

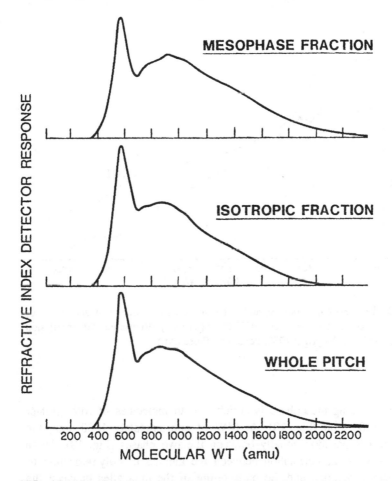

FIG. 12 Molecular weight distribution of the whole petroleum mesophase pitch and centrifuged phases from sample C in Table 2 (15 h at 400°C, 64.3% mesophase). (Reprinted with permission from Ref. 27, Copyright 1988, Pergamon Press Ltd.)

two phases. Since the coexisting isotropic phase contains mesogenic molecules as reported by Chen and Diefendorf [33], then the sizes of the mesogenic molecules in the isotropic phase are similar to the sizes found in the corresponding mesophase. The coexisting isotropic phase is enriched with molecules of MW less than 1000 amu, whereas the

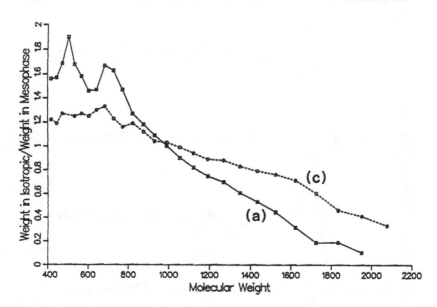

FIG. 13 Partition ratios of molecules in separated phases of sample A of Table 2 (curve a) and sample C of Table 2 (curve c). (Reprinted with permission from Ref. 27, Copyright 1988, Pergamon Press Ltd.)

corresponding mesophase is enriched with molecules of MW greater than 1000 amu (Fig. 13). At MW of approximately 1000 amu, the molecules partition equally between the two phases (Fig. 13). There exists a unique correlation between the kinetics of polymerization of a petroleum pitch and the partitioning of the molecules between the coexisting phases, since the reactivity of the pitch molecules becomes negligible at approximately the same MW (Fig. 8a) where the partition ratios equal unity (Fig. 13). This correlation between pitch reactivity and partitioning of molecules present in the coexisting phases will be used in a subsequent section to define the mechanism for mesophase sphere growth and mesophase development.

C. Characterization by Other Analytical Techniques

During the transformation of a petroleum pitch, the changes in the amount of short side chain protons (methyl groups) in the whole mesophase and in the separated phases have been reported by Greinke and Singer [27] and are also shown in Table 2. The percentages of short side chain protons in the whole pitch decreased slightly with time, and the percentages of short side chain protons in the mesophase were always less than those present in the corresponding isotropic phase. These results indicate that the methyl groups are concentrated in the lower MW species (monomers) and suggest that the pitch polymerization process that forms the dimers involves the cracking and loss of methyl groups.

Benn et al. [32] measured the glass transition temperature of centrifuged, separated phases in a polymerized petroleum ethylene cracker tar pitch and observed that the glass transition temperature of the mesophase did not change significantly until phase inversion of the emulsion is reached. The glass transition temperature of the isotropic liquid increases steadily during the transformation. The difference between the glass transition temperature of the coexisting phases in the early stages of mesophase growth is about 70°C, but this difference gradually decreases as the phase inversion region is approached. The result suggested that the changes in the apparent viscosity observed using in situ rheological measurements are largely due to changes in the viscosity of the isotropic phase and the changing volume fraction of mesophase at temperatures below the phase inversion point. This observation is consistent with the molecular weight distribution results of Fig. 11.

Singer et al. separated the coexisting phases in a petroleum mesophase pitch by high-temperature centrifugation and characterized them by determining their elemental constitutions, densities, solubilities, glass transition temperatures, molecular weight distributions, and x-ray and electron spin resonance (ESR) spectra [68]. The mesophase content in this pitch was high at 75%. The results tabulated in Table 4 show the extreme sensitivity of mesophase formation in petroleum pitches to the molecular structure and distribution. The relative density differences between the phases is about 4% for the petroleum pitch. The mesophase has a higher atomic C/H ratio than the coexisting isotropic phase. The higher C/H ratio and average molecular weight for the mesophase fraction (Fig. 11), as compared to the isotropic phase, is consistent with the role of aromatic dehydrogenative polymerization in the formation of mesophase in pitch [6]. The glass transition temperatures (T_g), measured by both differential scanning calorimetry (DSC) and thermal mechanical

TABLE 4   Characterization Data for the Separated Phases of a Petroleum
Mesophase Pitch Containing 75% Mesophase before Separation

	Petroleum mesophase	Petroleum isotropic
He density (g/cm^3)	1.32	1.27
% C	94.1	93.6
% H	4.43	4.74
Atomic C/H	1.78	1.66
% N	0.3	0.3
% S	1.15	1.17
Tg (°C) (DSC)	230	184
Tg (°C) (TMA)	225	185
% Carbon yield (1000°C)	73.6	65.6
X-ray interlayer distance (Å)	3.52	3.54
Spin concentration (10^{19}/g)	1.25	0.51
% Toluene insolubles	75	68
Molecular weight, ($\overline{M}_n$)	937	840

Reprinted in part with permission from Ref. 68, Gordon and Breach Science Publishers.

analysis (TMA), were about 40°C higher for the mesophase fraction. The
sulfur and nitrogen heteroatoms present in the petroleum pitch partitions
equally between the two phases. As deduced from Figs. 11 and 13, the
differences between many of these comparison values of Table 4 would
become smaller as the volume fraction of mesophase in the pitch
approached 100%. However, the differences between these comparison
values of Table 4 would become larger as the volume fraction of
mesophase in the pitch approached 0%.

The most surprising results obtained by the researchers using
centrifugation are the constant values of the constitution of the
mesophase during the transformation, as reflected in the constant values
of pyridine insolubles, glass transition temperatures, and molecular
weight distributions. How can this experimental observation be possible
since the formation of mesophase involves polymerization reactions?
Based on the previous information and quantitative analytical data
reviewed here regarding the kinetics of petroleum pitch polymerization,
an explanation can be derived and the mechanism of mesophase formation and sphere growth can now be described.

V. MECHANISM OF MESOPHASE FORMATION

A. Initial Mesophase Sphere Development

The nucleation of mesophase spheres in the isotropic pitch is due to the physical distillation of the pitch volatiles with MW less than 400 (Fig. 5) and the thermal polymerization of molecules in the 400–1100 MW range (Fig. 5). Both distillation and polymerization increase the average molecular weight of the isotropic phase and result in the initial development of long range molecular order in the pitch.

B. Additional Mesophase Development

Additional development of mesophase in mesophase containing devolatilized pitch should be separated into two parts.

1. In the coalesced mesophase and in the mesophase spheres, the 400–1100 MW molecules polymerize (Figs. 6a and 8a) causing a temporary increase in the average mesophase MW. The transfer of similar size molecules from the isotropic phase due to stirring and generated gas percolation restores molecular equilibrium in the mesophase. The transfer of these smaller pitch molecules, which are unreacted original pitch molecules, results in the growth of mesophase spheres and coalesced mesophase. The lack of reactivity of the largest molecules in the mesophase (Fig. 8a) also contributes to the constant molecular weight distribution and constant dispersion of the separated mesophase during the transformation (Table 3).

2. In the coexisting isotropic phase, the additional nucleation of mesophase spheres is due to two processes that increase the MW of the isotropic phase: One is the polymerization of the 400–1100 MW molecules present in the isotropic phase, whereas the other is the selective transfer of these similar size molecules from the isotropic phase to the mesophase, a process that restores molecular equilibrium in the mesophase. When the volume fraction of mesophase is small, sphere nucleation in the isotropic phase by polymerization of the 400–1100 MW predominates. However, when the volume fraction of mesophase is large, sphere nucleation in the isotropic phase by selective transfer of the 400–1100 MW molecules from the isotropic phase to the mesophase becomes dominant. When the isotropic phase is nearly depleted and the reservoir of smaller molecules in the isotropic phase becomes very small, the restoration of molecular equilibrium of the mesophase becomes difficult. Then the mesophase average MW

will begin to increase, and eventually the mesophase will become infusible. The glass transition temperature measurements of Benn et al. indicated that increases in glass transition temperature of the mesophase begins in an ethylene cracker pitch only as the phase inversion region is approached [32].

C. Other Considerations

Mochida et al. [45], in their early pioneering work on the structure of mesophase, suggested that sphere growth occurred by the incorporation of smaller molecules of the isotropic matrix into the spheres at a rate adequate to maximize their diameters. In contrast, however, Hüttinger and Wang [47] suggested that sphere growth occurred by the diffusion of mesogenic aromatics through the boundary layer. A mesogenic aromatic was defined as a large molecule thermally formed in the petroleum pitch, after distillation of the volatiles, from a sequence of several polymerization reactions. In this author's opinion, if sphere growth occurred by the assimilation of mesogens, which are defined as considerably higher molecular weight polymerization reaction products, one would have difficulty in maintaining the constant mesophase molecular weight distribution, PI content, and glass transition temperature during the transformation, especially since the smaller monomer molecules present in the mesophase continue to react.

The thermal conversion of petroleum pitch to mesophase involves both chemical polymerization and dealkylation reactions [26,61,69]. Hence, the development in the pitch of long-range molecular order has been attributed to both increasing molecular weight [26,61] and the loss of sp^3 type chemical bonding (side chains on aromatic rings) [69].

Such chemical reactions remove mesophase inhibitors and disordering species and lead to the formation of large planar molecules. The results of an analytical study [70] of the influence of side chains on the isotropic content in a decant oil-based mesophase pitch is shown in Fig. 14. Based on these data, it was subsequently calculated [70] that the dealkylation reactions, which increase the aromaticity and molecular planarity of the pitch molecules, have only a minor influence (approximately 3%) on mesophase formation, whereas polymerization reactions, which increase the molecular size, have a major influence (approximately 97%) on mesophase development. The relative importance of thermal polymerization and dealkylation reactions would vary somewhat based

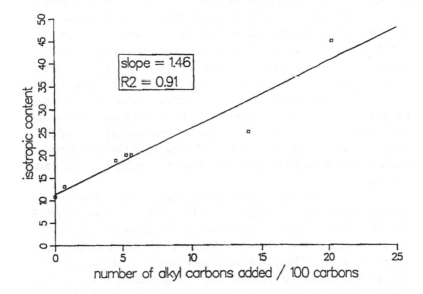

FIG. 14 Influence of alkyl carbon atoms added per average molecule via Friedel–Crafts reaction on the isotropic content in a petroleum mesophase pitch. (Reprinted with permission from Ref. 70, Copyright 1990, Pergamon Press Ltd.)

on the nature of the feedstock. For example, the contribution to mesophase development of dealkylation reactions occurring in coal tar pitches, which are more aromatic (fewer side chains) than decant oil pitches, would be even smaller.

D. Analytical Methods Based on Mesophase Formation Mechanisms

Several analytical methods that characterize mesophase pitch and use the principles of mesophase formation and support the characterization data described in the previous sections, have been developed. Lewis et al. [67] describe a quantitative determination of anisotropic domain size in mesophase pitch. The anisotropic domain size, defined as the average spacing between extinction contours observed during microscopic

examination with crossed polarizers, is known to be an important characteristic of mesophase pitch. For example, mesophase pitches employed
in the production of carbon fibers must be capable of forming a large
domain size when heated under quiescent conditions [71]. The average
domain size is related to the mesophase viscosity [72], which in turn is
related to the mesophase molecular weight distribution. The measured
domain size of the mesophase, obtained after heat treatment and annealing procedures, did not vary significantly for a given pitch as long as the
mesophase content was controlled between approximately 30 and 80%
[67]. The reason for the invariant mesophase domain size is the constant
molecular weight distribution of the mesophase during the transformation (Fig. 11).

Although numerous microscopy techniques have been developed for
quantifying mesophase content in pitch [73,74,75], a unique NMR
method for the quantitative determination of mesophase content of a
petroleum pitch was developed by Parks et al. [76]. A simple signal
analysis procedure allowed the anisotropic and isotropic phases of the
pitch to be distinguished quantitatively. The procedure is based on the
observation that the relaxation signals of protons present in the isotropic
phase and the mesophase decay with significantly different rates and can
readily be resolved graphically. The faster relaxing component of the
NMR signal was attributed to protons present in the mesophase. Parks
et al. observed a one-to-one correlation between the optically point-
counted mesophase (based on volume percent measurement and
measured at room temperature) and the residual NMR signal (measured
at 350°C). This observation implied that the hydrogen contents of
the mesophase and isotropic phases must be virtually the same. The
NMR data of Table 2 and the hydrogen content data of Table 4, however, show that the mesophase contains a slightly higher percentage
of aromatic protons and is slightly depleted in elemental hydrogen
compared to the isotropic phase owing to the dehydrogenative
polymerization reactions. The slightly higher density of the mesophase must exactly compensate for the slightly lower hydrogen concentration in the mesophase. Since the point count method is based
on a volume measurement, equal volumes of the coexisting phases
in the pitch, for example, result in essentially the same number of
protons in each phase. According to the data of Parks et al. [76],
this compensating condition must exist throughout the entire
transformation.

VI. PITCH POLYMERIZATION KINETICS AFTER MESOPHASE FORMATION

Previous sections of this review describe in detail the polymerization of a petroleum pitch during its transformation to the fluid mesophase. The next stage of the thermal carbonization process involves polymerization reactions of the mesophase constituents that significantly increase its viscosity and melting point to form a semicoke. A semicoke has been defined as a highly viscous mesophase produced from a pitch that has not yet advanced to infusible coke [77]. Intuitively, one would expect the kinetics and mechanisms of the petroleum pitch polymerization process to change at this point, since the previous kinetics results during the formation of mesophase indicated stability of the largest molecules or the dimers in the mesophase pitch (Fig. 8a).

When do these dimers begin to react? Greinke [26] has obtained the molecular weight distributions of a series of polymerized petroleum pitches, which have been subjected to heat treatment at 400°C for very long time periods between 33 and 263 h (Fig. 15). During this time period, the 100% mesophase pitch contains between 55 and nearly 100% quinoline insolubles (QI) content. The high QI contents suggest that semicokes or very high viscosity mesophase pitches with high softening points have been prepared from the petroleum pitch. The softening points of the pitches are higher than the temperature (400°C) at which the reaction kinetics were studied. Hence, the pitches are solids at this temperature. After 263 h of heat treatment, a peak appears at MW greater than 2000 amu. This peak is due to the exclusion of the 2000 and larger MW molecules from the pores of the GPC column. A smaller but still significant number of pitch molecules with MW between 400 and 600 remains in the semicoke even after 263 h of heating at 400°C. As a result, the number average molecular weight of the semicoke, heated for 263 h at 400°C, is only approximately 1300.

Visual evaluation of the curves of Fig. 15 reveal a slow but equal reactivity of the molecules with MW between 500 and 1200 amu. The polymerization of pitch molecules with MW greater than 1000 results in the buildup of molecules in the pitch with MW larger than 2000. The best fit first-order rate constants of the reaction of the petroleum pitch molecules shown in the sequence of Fig. 15 were reported [26] and are presented in Fig. 8b. The polymerization rate constants of the pitch molecules present in the advanced mesophase pitches or semicokes

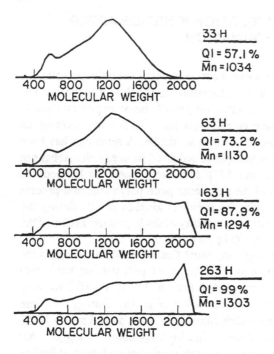

FIG. 15 Molecular weight distributions of a petroleum pitch after heat treatment for long times at 400°C. (Reprinted with permission from Ref. 26, Copyright 1986, Pergamon Press Ltd.)

(55–99% QI) at 400°C are now equal and are approximately seven times smaller when compared to the rate constants of the 400–800 MW molecules during mesophase formation (Figs. 6a and 8a).

Similarly, a first-order kinetic plot of the reaction of the methyl groups (quantified using ^{1}H-NMR and elemental hydrogen analysis) at 400°C in a petroleum pitch (Fig. 16) also show a significant decrease in the slope after the mesophase pitch is advanced to a very high viscosity [26]. Greinke noted that the ratio of the calculated rate constants for the loss of the methyl groups from petroleum pitch molecules during mesophase formation and after thermal advancement of the mesophase is approximately seven, a value similar to the ratio of rate constants previously measured from the GPC curves during the two carbonization regimes (Fig. 8). The correlation between the GPC kinetics and the ^{1}H-NMR kinetics suggests

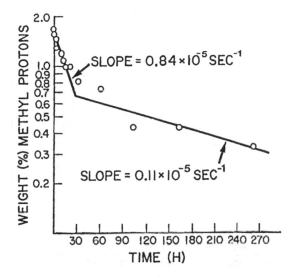

FIG. 16 First-order kinetic plot of the reaction of the methyl protons in a petroleum pitch as a function of time at 400°C. (Reprinted with permission from Ref. 26, Copyright 1986, Pergamon Press Ltd.)

that methyl groups of a petroleum pitch are deeply involved in the polymerization process through the semicoke stage [26]. The significant change in the kinetics of polymerization after advancement of the meso-phase was shown to be due to a physical phenomenon in which solid-state reaction kinetics are different from the liquid-state kinetics [26].

The altering of the kinetics after thermal advancement of the mesophase is not unusual. The effect is identical to reactions in other thermosetting systems, whose reactions also proceed much more slowly after vitrification [78,79,80]. Gillham defines vitrification of thermo-setting systems to occur when the glass transition temperature increases to the isothermal temperature of cure [78]. Vitrification results in the quenching of chemical reactions.

The significant drop in the reaction kinetics after vitrification of the petroleum pitch (Figs. 8 and 16) is consistent with the gassing results of a coal tar pitch obtained by Politis and Chang [81]. Based on the data of Figs. 8 and 16, one would predict a significant and abrupt reduc-tion of carbonization gases after mesophase is thermally advanced. The

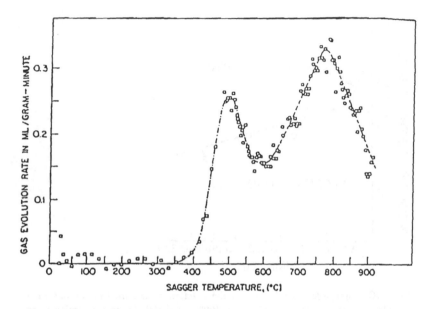

FIG. 17 Rate of gas evolution from a coal tar pitch as a function of heat treatment temperature. (Reprinted with permission from the authors of Ref. 81.)

nonisothermal gas evolution work of Politis and Chang (Fig. 17) shows two gassing peaks with the first occurring at approximately 460°C and the second at 730°C. The first maximum gassing peak occurs when the mesophase is still fluid. Most of these evolved gases generated during this peak are due to the polymerization reactions of the pitch monomers (Fig. 5). These gas-generating reactions involve the formation of an aryl–aryl oligomer through combination of the initial rapid reaction of the sigma free radicals [57]. The minimum gassing rate at 580°C occurs when the viscosity of the mesophase pitch has increased significantly and the rates of the carbonization reactions have decreased considerably as observed in Figs. 8 and 16. The second maximum peak at 730°C is now due to the carbonization gases evolved from the polymerization and condensation reactions after the pitch has vitrified and formed an infusible coke. Many of the stable pi free radicals, as observed by electron spin resonance, are formed during this gassing regime as a result of the dehydrogenative condensation reactions [57].

VII. CONCLUDING REMARKS

The application of special procedures to the conventional analytical techniques of gel permeation chromatography, centrifugation, and nuclear magnetic resonance has provided unique insights pertaining to the chemistry, kinetics, and mesophase-forming mechanisms of petroleum pitches. Continued research to clarify our basic understanding of the mechanisms of carbonization in practical raw materials, such as petroleum pitch, is essential in order to attain the maximum potential use of these materials for producing carbon and graphite products.

ACKNOWLEDGMENTS

The author thanks Dr. L. S. Singer and Ms. J. M. Greinke for their helpful criticism of the manuscript. This review is based on a seminar presented to the Fuel Science Program at The Pennsylvania State University in April 1991.

REFERENCES

1. H. Marsh (ed.), *Introduction to Carbon Science*, Butterworth, London, 1989.
2. I. C. Lewis, in *Encyclopedia of Chemical Technology*, Fourth Edition, Vol. 4 (I. Kroschwitz and M. Howe-Grant, eds.), Wiley, New York, 1992, p. 953
3. M. Zander, *Fuel 66*, 1459 (1987).
4. I. C. Lewis, *Carbon 20*, 519 (1982).
5. J. D. Brooks and G. H. Taylor, *Nature 206*, 697 (1965).
6. L. S. Singer, *Carbon 16*, 408 (1978).
7. R. T. Lewis and I. C. Lewis, *Carbon 26*, 757 (1988).
8. H.-G. Franck and J. W. Stadelhofer, *Industrial Aromatic Chemistry*, Springer-Verlag, New York, 1987, p. 63.
9. H. E. Romine and J. R. McConaghy, Jr., U.S. Patent 5,032,250 (July 16, 1991).
10. K.-H. Kochling, B. McEnaney, S. Muller, and E. Fitzer, *Carbon 23*, 601 (1985).
11. I. Mochida, Y. Korai, A. Azuma, M. Kakuta, and E. Kitajima, *J. Mat. Sci. 26*, 4836 (1991).
12. M. Zander, *Fuel 66*, 1536 (1987).

13. R. A. Forrest, H. Marsh, and C. Cornford, in *Chemistry and Physics of Carbon*, Vol. 19 (P. A. Thrower, ed.), Marcel Dekker, New York, 1984, p. 211.
14. B. Rand, "Carbon Fibres from Mesophase Pitch," in *Handbook of Composites*, Vol. 1 (W. Watt and B. V. Perov, eds.), Elsevier Science, New York, 1985, p. 495.
15. T. Edstrom and B. A. Petro, *J. Poly Sci., Part C 21*, 171 (1968).
16. I. C. Lewis and B. A. Petro, *J. Poly. Sci., Polymer Chem. Edn. 14*, 1975 (1976).
17. R. A. Greinke and I. C. Lewis, *Anal. Chem. 47*, 2151 (1975).
18. E. M. Dickinson, *Fuel 59*, 290 (1980).
19. I. G. Bergmann, L. J. Duffy, and R. B. Stevenson, *Anal. Chem. 43*, 131 (1971).
20. M. Popl, J. Fahnrich, and M. Stejskal, *J. Chromatogr. Sci. 14*, 537 (1976).
21. H. H. Oelert, D. R. Latham, and W. E. Haines, *Sep. Sci. 5*, 657 (1970).
22. F. H. Tillmann, W. Ulsomer, and G. Piezka, Preprints, Carbon 76, 2nd International Carbon Conference, 1976, p. 385.
23. L. Reggel, R. A. Friedel, and I. Wender, *J. Org. Chem. 22*, 891 (1957).
24. J. D. Brooks and H. Silberman, *Fuel 41*, 67 (1962).
25. H. W. Sternberg, C. L. Delle Donne, P. Dantages, E. C. Moroni, and R. E. Markby, *Fuel 50*, 732 (1971).
26. R. A. Greinke, *Carbon 24*, 677 (1986).
27. R. A. Greinke and L. S. Singer, *Carbon 26*, 665 (1988).
28. R. A. Greinke and L. H. O'Connor, *Anal. Chem. 52*, 1877 (1980).
29. W. W. Yau, J. J. Kirkland, D. D. Bly, and J. J. Stocklosa, *J. Chromatogr. 125*, 219 (1976).
30. J. D. Brooks and G. H. Taylor, *Carbon 3*, 185 (1965).
31. B. Rand and S. Whitehouse, *Ext. Abstracts 16th Biennial Conference on Carbon*, American Carbon Society, San Diego, California, 1983, p. 102.
32. M. Benn, B. Rand, and S. Whitehouse, *Ext. Abstracts 17th Biennial Conference on Carbon*, American Carbon Society, Lexington, Kentucky, 1985, p. 159.
33. S. H. Chen and J. J. Diefendorf, *Ext. Abstracts, Carbon 84, International Carbon Conference*, Bordeaux, France, 1984, p. 382.
34. L. S. Singer, D. M. Riffle, and A. R. Cherry, *Carbon 25*, 249 (1987).

35. C. I. Grindstaff, L. A. Bryan, and M. P. Whittacker, Preprints, Division of Petroleum Chemistry, Am. Chem. Soc., New York Meeting, August 27, 1972, p. 115.
36. H. L. Retcofsky and R. A. Friedel, *Spectrometry of Fuels* (R. A. Friedel, ed.), Plenum Press, New York, 1970, Chapter 6.
37. R. A. Greinke, *Fuel 63*, 1374 (1984).
38. K. Miyazawa, T. Yokono, Y. Sanada, E. Yamada, and S. Shimokawa, *Carbon 19*, 143 (1981).
39. T. Yokono, K. Miyazawa, Y. Sanada, and S. Shimokawa, *Fuel 58*, 239 (1979).
40. A. Kiyoshi, T. Yokono, Y. Sanada, and S. Uemura, *Ext. Abstracts, 17th Biennial Conference on Carbon*, American Carbon Society, Lexington, Kentucky, 1985, p. 157.
41. M. Hamaguchi and T. Nishizawa, *Fuel 71*, 747 (1992).
42. T. Nishizawa and M. Sakata, *Fuel 70*, 124 (1991).
43. V. L. Weinberg, B. C. Gerstein, P. D. Murphy, and T. F. Yen, *Ext. Abstracts 15th Biennial Conference on Carbon*, American Carbon Society, Philadelphia, Pennsylvania, 1981, p. 132.
44. I. Mochida, Y. Korai, A. Azuma, M. Kakuta, and E. Kitajima, *J. Mat. Sci. 26*, 4836 (1991).
45. I. Mochida, K. Maeda, and K. Takeshita, *Carbon 15*, 17 (1977).
46. I. Mochida, K. Maeda, and K. Takeshita, *Carbon 16*, 459 (1978).
47. K. J. Hüttinger and J. P. Wang, *Carbon 30*, 1 (1992).
48. P. Delhaes, J. C. Rouillon, G. Fug, and L. S. Singer, *Carbon 17*, 435 (1975).
49. R. A. Greinke, *Carbon 30*, 407 (1992).
50. H. Marsh and P. L. Walker, Jr., *Chemistry and Physics of Carbon*, Vol. 15 (P. L. Walker, Jr., and P. A. Thrower, eds.) Marcel Dekker, New York, 1979, p. 229.
51. I. C. Lewis and L. S. Singer, *Ext. Abstracts 12th Biennial Conference on Carbon*, American Carbon Society, Pittsburgh, Pennsylvania, 1975, p. 265.
52. H. Honda, M. Kimura, Y. Savada, S. Sugawara, and T. Faruta, *Carbon 8*, 181 (1970).
53. L. S. Singer and I. C. Lewis, *Carbon 16*, 417 (1978).
54. J. D. Brooks and G. H. Taylor, *Chemistry and Physics of Carbon*, Vol. 4 (P. L. Walker, Jr., ed.) Marcel Dekker, New York, 1968, p. 271.
55. I. C. Lewis and R. Didchenko, Preprints, 3rd International Carbon Conference, Baden-Baden, Germany, 1976, p. 385.

56. Y. Yamada, S. Oi, H. Tsutsui, and E. Kitajima, *Ext. Abstracts 12th Biennial Conference on Carbon*, American Carbon Society, Pittsburgh, Pennsylvania, 1975, p. 271.

57. I. C. Lewis and L. S. Singer, *Chemistry and Physics of Carbon*, Vol. 17 (P. L. Walker, Jr., and P. A. Thrower, eds.) Marcel Dekker, New York, 1981, p. 1.

58. R. H. Schlosberg, M. C. Gorbaty, and T. Aczel, *J. Amer. Chem. Soc.* *100*, 4188 (1978).

59. M. Farcasiu, T. G. Mitchell, and D. D. Whitehurst, Abstracts, 172nd National Meeting of the American Chemical Society, Fuel Division, *21*, 11 (1976).

60. L. R. Rudnick and D. R. Tueting, *Ext. Abstracts 18th Biennial Conference on Carbon*, American Carbon Society, Worcester, Massachusetts, 1987, p. 66.

61. R. A. Greinke and I. C. Lewis, *Carbon 22*, 305 (1984).

62. I. C. Lewis and A. W. Moore, U.S. Pat. No. 4,317,809 (1982).

63. Y. D. Park and I. Mochida, *Carbon 27*, 925 (1989).

64. R. T. Lewis and I. C. Lewis, US Patent 4,303,631 (December 1, 1981).

65. S. Chwastiak and I. C. Lewis, *Carbon 16*, 156 (1978).

66. R. T. Lewis, *Ext. Abstracts 12th Biennial Conference on Carbon*, American Carbon Conference, Pittsburgh, Pennsylvania, 1975, p. 215.

67. R. T. Lewis, I. C. Lewis, R. A. Greinke, and S. L. Strong, *Carbon 25*, 289 (1987).

68. L. S. Singer, I. C. Lewis, and R. A. Greinke, *Mol. Cryst. Liq. Cryst.* *132*, 65 (1986).

69. D. M. Riggs and R. J. Diefendorf, *Ext. Abstracts 14th Biennial Conference on Carbon*, American Carbon Conference, University Park, Pennsylvania, 1979, p. 413.

70. R. A. Greinke, *Carbon 28*, 701 (1990).

71. L. S. Singer, U.S. Patent 4,005,183 (January 25, 1977).

72. J. D. Brooks and G. H. Taylor, *Chemistry and Physics of Carbon*, Vol. 4 (P. L. Walker, Jr., ed.) Marcel Dekker, New York, 1968, p. 243.

73. S. Chwastiak, R. T. Lewis, and J. D. Ruggiero, *Carbon 19*, 357 (1981).

74. D. R. Ball, *Ext. Abstracts 18th Biennial Conference on Carbon*, American Carbon Society, Worcester, Massachusetts, 1987, p. 173. ASTM Method D4616-87.

75. H. Sunago, M. Higuchi, and T. Tomioka, *Ext. Abstracts 19th Biennial Conference on Carbon*, American Carbon Society, University Park, Pennsylvania, 1989, p. 102.
76. T. J. Parks, L. F. Cross, and L. J. Lynch, *Carbon 29*, 921 (1991).
77. L. S. Singer and D. T. Orient, U.S. Patent 4,891,203 (January 2, 1990).
78. J. K. Gillham, *Proceedings of the 13th North American Thermal Analysis Society Conference*, Philadelphia, 1984, p. 344.
79. R. B. Prime, in *Thermal Characterization of Polymeric Materials* (Edith A. Turi, ed.), Academic Press, New York, 1981, p. 435.
80. J. K. Gillham, "The Time-Temperature-Transformation (TTT) State Diagram and Cure," in *The Role of the Polymer Matrix in the Processing and Structural Properties of Composite Materials* (J. C. Seferis and L. Nicolais, eds.), Plenum Press, New York, 1983, pp. 127–145.
81. T. G. Politis and C. F. Chang, *Ext. Abstracts 17th Conference on Carbon*, American Carbon Society, Lexington, Kentucky, 1985, p. 8.
82. I. C. Lewis, *J. Chim. Phys. 81*, 751 (1984).

2

Thermal Conductivity of Diamond

Donald T. Morelli

General Motors Research and Development Center, Warren, Michigan

I. INTRODUCTION

Harder than steel. More thermally conducting than copper. Chemically inert to everything below 700 K. As slippery as Teflon. More electrically resistive than any element. Among the rarest of gemstones.

The possession of any one of these properties is reason enough to make a material unique. That they are features of a single substance has made the form of carbon known as diamond one of the most prized materials known to man. Paradoxically, nature in the extreme has made diamond the most useful and the most useless material at the same time. Rare in the wild and stubborn to tame synthetically in the laboratory, diamond possesses yet another extreme—its cost—which has prevented the exploitation of its extreme physical properties on a widespread basis in industry and technology.

And yet, now, we find ourselves in the midst of nothing less than a diamond revolution. Propelled by rapid advances in the nonequilibrium growth from the vapor phase [1] hopes have been rekindled that mass production of a high-quality single-crystal diamond will at last become a reality.

This potential breakthrough in diamond technology has spurred researchers to examine more closely this material in both its natural and synthetic forms. One very important physical property of diamond, which has been a major area of interest, is its thermal conductivity. Carrying heat a full factor of five better than copper at room temperature, single-crystal diamond is the best heat conductor known to man. The use of diamond as a heat-spreading or heat transfer medium in electronic device applications has thus received a great deal of attention [2].

Because the Debye temperature of diamond is very large (~2000 K), intrinsic phonon–phonon scattering is fairly weak at room temperature compared to other crystalline materials. As a result, minute levels of impurities can play a dramatic role in limiting the thermal conductivity of both natural single crystals and man-made films over a surprisingly wide temperature range. A measure of the thermal conductivity not only provides important data for applications, but, when carried out as a function of temperature, is also a powerful method of characterizing the defect state of a crystal. In this review I wish to summarize the experimental situation with regard to the thermal conductivity of diamond. My main focus here is on single crystals and chemically vapor-deposited films, since these represent the purest forms of this material. I will not discuss thermal measurements on diamond sinters, powders, or composites of diamond and other materials. Although these substances are important in their own right, they are best left to separate consideration. Blanchard et al. [3] have recently surveyed experimental results on the thermal conductivity of single-crystal and polycrystalline diamond. For the most part, I will emphasize those measurements made as a function of temperature because they contain the most valuable information about the intrinsic and extrinsic scattering mechanisms in this material; I will, however, make reference to other measurements made at a single temperature when they are appropriate to the discussion at hand.

Section II of this review gives a very brief outline of the theory of the thermal conductivity of nonmetallic solids. In Section III we will consider the experimental information on the thermal conductivity of diamond single crystals, whereas Section IV does the same for chemically vapor-deposited films. Finally, Section V provides a summary and some brief remarks about future directions of research in this area.

II. THEORETICAL BACKGROUND

A. Introduction

With a band gap of 5.4 eV, diamond is classified as a wide-band-gap semiconductor. As such, even at temperatures as high as the Debye temperature, there are few free electrons to carry heat and contribute to the thermal conductivity. Thus, virtually all the heat is carried by lattice vibrations, or phonons. The thermal conductivity of nonmetallic crystals has been the subject of several reviews and monographs [4–6] in which the interested reader will find exhaustive details of both theory and experiment. What follows here is a brief overview of the subject, sufficient for the discussion at hand.

The flow of heat in a nonmetallic crystal such as diamond implies that
the phonon distribution differs from that under conditions of thermal
equilibrium. From the point of view of understanding experimental
thermal conductivity data, the degree of departure of the phonon distribution from equilibrium is most conveniently expressed in terms of relaxation times τ. In this *relaxation-time approximation*, various phonon
modes are assumed to be scattered by several processes that in general
depend on phonon polarization and frequency, as well as temperature.
The thermal conductivity can then be expressed as [4]

$$\kappa = 1/3 \int_0^{\omega_{max}} \hbar \, \omega v^2 \tau f(\omega) dN_0/dT \; d\omega, \tag{1}$$

where $2\pi\hbar$ is the Planck constant; ω_{max}, the maximum phonon frequency; v, the phonon group velocity; $f(\omega)d\omega$, the number of phonons
between ω and $\omega + d\omega$; and N_0, the phonon distribution function. Two
further assumptions need to be made in order to cast the thermal conductivity into a form more easily comparable with experiment: (1) a linear
dispersion relation $\omega = vq$ between phonon frequency and wavevector, q
and (2) the equality of phonon velocities for all polarizations. This leads
to the expression

$$\kappa = (1/2\pi^2 v) \int_0^{\omega_{max}} \hbar\omega^3 \tau (\hbar\omega/kT^2) \exp(\hbar\omega/kT) \, [\exp(\hbar\omega/kT)-1]^{-2} \, d\omega \tag{2}$$

where k is the Boltzmann constant. By a simple change in variable
$x=\hbar\omega/k_B T$, this equation becomes

$$\kappa = (k/2\pi^2 v) \, (k/\hbar)^3 T^3 \int_0^{\theta/T} [\tau(x) e^x/(e^x - 1)^2] dx \tag{3}$$

where $\theta = \hbar\omega_{max}/k$ is the Debye temperature. One can immediately
recognize that this expression can be cast in terms of the spectral specific
heat $C(x)$ as

$$\kappa = 1/3 v^2 \int_0^{\theta/T} \tau(x) C(x) dx = 1/3v \int_0^{\theta/T} l(x) C(x) dx \tag{4}$$

where $l(x)$ is the frequency-dependent phonon mean free path. This
formula is analogous to the kinetic expression $\kappa = Cvl$ for a gas, the factor
of 1/3 coming from averaging over three phonon modes. Figure 1 shows
the spectral specific heat $C(x)$. It can be seen that, even though all

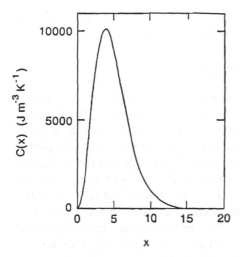

FIG. 1 Spectral specific heat of diamond at 100 K, using a photon velocity $v = 1.32 \times 10^4$ m s^{-1}, a Debye temperature $\theta = 2000$ K.

phonons up to the maximum frequency contribute to $C(x)$, the spectral specific heat has a strong peak at $\omega=3.8k_BT/\hbar$; phonons of this frequency are called *dominant phonons* and are responsible for carrying the majority of heat at the given temperatures.

The problem now becomes one of determining the proper form of the total phonon scattering time $\tau(\omega)$, which is the inverse of the sum of the scattering *rates* of the various scattering processes. Although in principle this is straightforward, in practice there are a number of circumstances, one example of which is crystalline anisotropy, that can complicate the situation. Nonetheless Eq. (2) can be quite useful since $\tau(\omega)$ will have different frequency dependences for different scattering mechanisms, and these give rise to characteristic temperature dependences of κ. Thus, simple examination of the temperature dependence of the thermal conductivity can yield valuable qualitative information about the dominant scattering processes in a particular crystal.

In order to facilitate a comparison of (2) with experiment, one needs to know the form of the various scattering rates that comprise τ. I shall consider three main types of phonon scattering in a nonmetallic crystal: (1) intrinsic phonon–phonon processes, (2) phonon-defect and phonon-impurity interactions, and (3) boundary scattering. I will now briefly discuss each of these mechanisms in detail.

B. Intrinsic Phonon–Phonon Interactions

In 1911 Eucken [7] noted that in most dielectric materials the thermal conductivity was proportional to $1/T$ above about 80 K. This was explained by Debye [8], who pointed out that local fluctuations in the density of the lattice cause scattering of lattice waves, which, assuming a constant specific heat, give rise to a scattering time and thus a thermal conductivity inversely proportional to T. The first rigorous treatment of the problem was given by Peierls [9]. He showed that if the potential energy of an atom in a lattice were an exact quadratic function of displacement from equilibrium, then no interactions between phonons would occur, and, in the absence of any other scattering processes, the thermal conductivity would be infinite. *Phonon–phonon interactions* arise only because of anharmonic coupling between lattice waves. For phonons of wavevector q, Peierls further considered two types of three-phonon interactions of the form

$$q_1 + q_2 = q_3 + Q \tag{5}$$

namely, those that conserve crystal momentum, called *normal*, or n-processes, for which $Q = 0$, and those that do not conserve crystal momentum, called umklapp, or u-, processes, which involve the addition of a nonzero reciprocal lattice vector Q. Because normal processes do not degrade the total momentum of the phonon system, rather simply redistribute it amongst the phonons, they do not directly give rise to thermal resistance (this is not to say that normal processes play no role in the thermal resistivity; more about that follows). Only umklapp processes contribute directly to the thermal resistance. Peierls showed that at high temperature ($T > \theta$) this resistance is proportional to T but decreases exponentially at low temperatures due to the rapidly decreasing availability of phonons that can satisfy (5) with $Q \neq 0$.

An estimate of the magnitude of the thermal resistance due to u-processes was given by Leibfried and Schloemann [10], who obtained

$$\kappa = (B/\gamma)(k_B/h)^3 Ma\theta^3/T, \tag{6}$$

where γ is the Grueneisen parameter, which is a measure of the anharmonicity of the lattice; M, the atomic mass; a^3, the atomic volume; and

$B \simeq 3.8$. The same expression but with different values of B has been given by Julian [11] and Klemens [12]. Slack [13] modified this formula to include materials with more than one atom per unit cell and showed that most nonmetallic crystals did indeed roughly agree in magnitude with its predictions.

Even though the $1/T$ law for $T > \theta$ is reasonably well established both theoretically and experimentally, the exact form of the u-process scattering rate and its dependence on frequency and temperature for a given material is not. It turns out that the calculation of such a rate represents a formidable theoretical problem. As a result, a wide range of forms for this scattering rate have been introduced on an empirical basis in order to fit thermal conductivity data. Some guidance is provided by Herring [14], who suggests that for phonon–phonon interactions,

$$1/\tau \sim \omega^a T^{5-a} \exp(-\theta/bT) \tag{7}$$

and provides values of a for various crystal structures. For the cases of silicon and germanium, for instance, Glassbrenner and Slack [15] found good agreement with the experimental data up to 1000 K for $a = 2$ (in agreement with Herring's prediction) and $b \simeq 3$. This example is particularly relevant because the range of temperature of the experimental data extend well above the Debye temperature. Many other forms of Eq. (7) have been used for various materials, and, indeed, when the experimental data cover a fairly narrow range in temperature of the umklapp regime, it is difficult to distinguish between fits to the thermal conductivity data using different values of a and b. Extreme care must be exercised when attaching physical significance to the chosen values.

Even though *normal three-phonon processes* do not directly contribute to the thermal resistance because they do not degrade the total phonon momentum, they can have a profound effect under certain circumstances. Basically, this is because a normal phonon scattering process can generate a phonon that is more likely to be umklapp scattered or scattered by a defect and thereby forfeit some of its momentum. The most widely accepted method for treating the effects of normal processes was given by Callaway [16]. He showed that if the normal phonon scattering time is τ_n and umklapp scattering time is τ_u (with the total phonon scattering rate given by $\tau_c{-}1 = \tau_n{-}1 + \tau_u{-}1$), then the conductivity in the relaxation time approximation is given by

$$\kappa = \kappa_1 + \kappa_2$$

$$\kappa_1 = (k/2\pi^2 v)\,(k/\hbar)^3 T^3 \int_0^{\theta/T} \tau_c x^4 e^x (e^x - 1)^{-2}\, dx$$

$$\kappa_2 = (k/2\pi^2 v)\,(k/\hbar)^3 T^3\, [\int_0^{\theta/T} (\tau_c/\tau_n) x^4 e^x (e^x - 1)^{-2}\, dx]^2 / \qquad (8)$$

$$\int_0^{\theta/T} (\tau_c/\tau_n \tau_u) x^4 e^x (e^x - 1)^{-2}\, dx$$

Thus, even though normal processes are not a direct source of thermal resistance, the normal scattering time τ_n appears in this expression and in some circumstances will play a role in determining the nature of the thermal conductivity. Berman [4] discusses in detail the conditions under which normal processes are expected to be important. Generally, in crystals that contain moderate amounts of defects or several isotopes, normal processes can usually be neglected. In a very pure crystal or one containing only a small concentration of an impurity isotope, however, normal processes can contribute and may even be the dominant process determining the thermal resistivity. As in the case of umklapp scattering, the exact dependence of the normal phonon-scattering rate on frequency and temperature is in most cases incalculable, and one must again resort to an empirical expression when attempting to fit experimental data.

Even though Peierls predicted that for $T < \theta$ the thermal conductivity should increase exponentially with falling temperature, such behavior was not found, except in a few isolated cases such as sapphire [17], titanium oxide [18], and bismuth [19]. In many cases the experimental results were explained as being due to the presence of defects in the crystals, but even some materials that were apparently defect-free, such as silicon and germanium [20], did not exhibit the predicted temperature dependence. It was soon recognized [21], however, that even in pure crystals, isotopes would provide local mass fluctuations that would lead to additional scattering of phonons. This was an old idea dating back to Pomeranchuk [22]. The scattering of lattice waves by mass and elastic constant differences thus became a central focus of theoretical calculations.

C. Phonon-Defect Interactions

The case of an isotope of an atom in the lattice is a special case of a more general class of defects called *point defects*. In general, a point defect is any defect of spatial dimension smaller than the phonon wavelength.

Since at any given temperature phonons of all frequencies up to the Debye frequency are available, a defect is or is not a point defect depending on the wavelength of the particular phonon with which is interacts. Examples of point defects are substitutional and interstitial impurities and vacancies. Scattering arises due to a difference in mass and/or elastic constants in the neighborhood of the impurity. In the case of an isotope, there is only a mass difference, and for this type of impurity, Klemens [23] derived the mass-difference scattering rate

$$(\tau_i)^{-1} = (ca^3/4\pi v^3)\omega^4(\Delta M/M)^2 \tag{9}$$

where c is the fractional concentration of isotope and ΔM is its difference in mass with respect to the average atomic mass. In the more general case of an impurity atom of different size or a vacancy, there is, in addition to the mass difference, a difference in linkages between the impurity and the neighboring atoms. For dilute concentrations of impurities, this gives rise to a point-defect scattering rate of approximately the form

$$(\tau_{pd})^{-1} = (ca^3/4\pi v^3)\omega^4[\Delta M/M + 2\Delta\zeta/\zeta] \tag{10}$$

where ζ is the host atomic force constant and $\Delta\zeta$ is the difference between the impurity–host and host–host force constants.

An examination of the frequency dependence of the point-defect scattering rate indicates that this type of scattering will be important at high temperatures when the dominant phonons have high frequencies and will fall off strongly for low temperatures and small phonon frequencies. In fact, if one calculates the thermal conductivity from Eq. (2) using only umklapp and point-defect scattering, one finds that κ diverges at low temperatures. Physically, due to the rapid disappearance of u-processes and the ω^4 falloff of the point-defect scattering rate, there is nothing to prevent the long wavelength phonons from running away with the heat without being scattered, a concept is intuitively unacceptable even for a perfect crystal. Ultimately, however, the phonon mean free path is limited by scattering from the surfaces of the crystal, and this mechanism provides a limit to the thermal conductivity at low temperatures. This will be discussed more thoroughly in the next subsection.

Positive evidence of the isotope effect on the thermal conductivity was provided by Geballe and Hull [24], who measured the thermal conductivity of two single crystals of germanium, one of natural isotopic composition and one that had undergone isotopic purification. The enhancement in conductivity at low temperatures in the purified crystal was found to bc in quite satisfactory agreement with the theory of isotope resistance [16]. Quantitative comparison between the point-defect scattering rate formula and experimental data for impurity atoms of different size and for vacancies has not always been so successful, however, mainly because of the difficulty in estimating the degree of strain in the lattice caused by the presence of the impurity.

Because at low temperatures the point-defect scattering rate becomes small and at high temperatures it becomes overshadowed by intrinsic phonon–phonon scattering, point-defect scattering tends to influence the thermal conductivity most dramatically at the peak in the thermal conductivity. A beautiful example of isotope scattering was provided by Berman and Brock [25], who studied LiF compounds for a range of ^{6}Li concentrations. Their results are shown in Fig. 2.

Even though point defects are probably the most prevalent type of defect in a crystal and the most straightforward to compare with

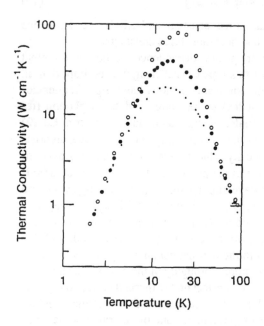

FIG. 2 Thermal conductivity of LiF for the following ^{6}Li concentrations: (O), 0.01%; (●), 4.6%; (·), 50.1%. Adapted from Ref. 25.

theoretical predictions, there are many other types of defects that can scatter phonons and thereby degrade the thermal conductivity. Only a few other examples will be given here. The strain field about a dislocation will scatter phonons at a rate given by [5]

$$(\tau_d)^{-1} = 0.06 N_d b^2 \gamma^2 \omega \qquad (11)$$

where N_d is the number of dislocations per unit area and b is the Burgers vector. This type of *dislocation scattering* results in a thermal conductivity (in the absence of any other scattering process) proportional to T^2. An example of dislocation scattering in a crystal into which a large density of dislocations has been purposely introduced [26] is shown in Fig. 3. Generally, there has not been a great deal of success in comparing the dislocation density determined from a fit to the thermal conductivity with that determined by other means, such as etch pit studies. Usually the number of dislocations required to fit the experimental data exceeds

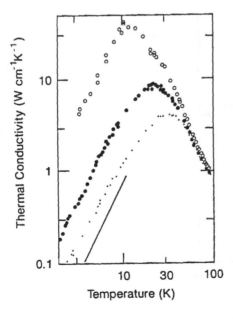

FIG. 3 Thermal conductivity of LiF with various dislocation densities induced by compressive deformation: (○), undeformed crystal; (●), crystal reduced in length 2.4%; (·), crystal reduced in length 4.0%. After Ref. 26.

the number determined from microscopy. Various explanations such as mobile [27] or vibrating [28] dislocations have been presented to explain these discrepancies.

Some defects, such as molecular impurities and certain vacancies, give rise to phonon scattering, which can be described by a *resonant scattering* rate given by [29]

$$(\tau_r)^{-1} = A\omega^2/(\omega^2 - \omega_0^2)^2 \tag{12}$$

where A is proportional to the concentration of resonant scatterers. The thermal conductivity under these circumstances shows a characteristic dip centered at a temperature corresponding to the dominant phonon frequency equal to ω_0. An example of this type of scattering is shown in Fig. 4.

Another situation that can produce a dip in the thermal conductivity curve such as that due to resonant scattering occurs in the presence of

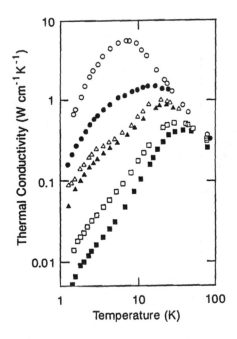

FIG. 4 Thermal conductivity of KCl containing NO_2^- ions according to Ref. 29. Symbol designation: (O), pure crystal; ($\bullet$), 9×10^{16} cm^{-3}; ($\triangle$), 4×10^{17} cm^{-3}; ($\blacktriangle$), 5×10^{17} cm^{-3}; ($\square$), 1.6×10^{18} cm^{-3}; ($\blacksquare$), 4×10^{18} cm^{-3}.

large-scale impurity aggregates or precipitates of spatial dimension L. In this case, the defect will behave as a point defect for phonon wavelengths larger than L and as a geometrical obstruction for short phonon wavelengths. In particular, the *precipitate scattering* rate is of the form [30]

$$(\tau_p)^{-1} = C\omega^4 \text{ for } qa < 1$$
$$= \text{const for } qa > 1 \tag{13}$$

Thus, at high temperatures, the scattering is constant, but falls off as ω^4 at temperatures below the crossover set by the condition $q_{dom}a=1$ when the dominant phonon wavevector exceeds $1/a$. An example of this type of scattering for the case of divalent impurities in KCl [30] is shown in Fig. 5. The similarity between the shapes of the conductivity curves shown in Figs. 4 and 5 means one must be very careful in attributing the dips to either a resonant or precipitated defect, especially in highly defective materials when both types of defect are likely to occur.

It should be clear from the preceding discussion that, due to various frequency dependences of phonon-defect scattering mechanisms, one can use a measurement of the thermal conductivity of a nonmetallic solid as a sort of crude but effective "phonon spectroscopy" to determine the nature, size, and concentration of defects in these materials. Standing alone this technique is not very powerful, but when used in conjunction with other more common measurements, particularly optical studies, it can provide important information not readily obtainable by other means.

D. Phonon-Boundary Scattering

As mentioned in the previous section, at low temperatures in a pure crystal or a crystal containing only point defects, the phonon mean free path increases strongly with decreasing temperature. This would seem to imply that as $T \to 0$ the thermal conductivity would diverge because there are no scattering processes to prevent long wavelength phonons from running away with the heat. When the phonon mean free path becomes on the order of the sample size (for a single crystal) or the crystallite site (for a polycrystal), phonons scatter off of the boundaries, which leads to a nonzero thermal resistance. *Boundary scattering* such as that just described was first suggested by Casimir in 1938 [31]. He showed that, if it is assumed that the scattering of a phonon from a crystal

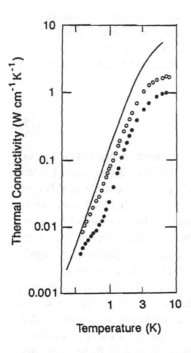

FIG. 5 Thermal conductivity of KCl containing divalent impurities: solid line, pure crystal; (O), KCl containing 5.8×10^{18} cm^{-3} Ba^{2+}; (●), KCl containing 5.4 $\times 10^{18}$ cm^{-3} Sr^{2+}. Adapted from Ref. 30.

boundary is completely diffuse, the phonon mean free path would be limited to a value roughly equal to the smallest dimension of the sample (or crystallite). In particular, for a crystal of square cross section d, the diffuse phonon-boundary scattering rate is given by

$$(\tau_b)^{-1} = v/1.12d \tag{14}$$

Note that this process is both temperature and frequency independent; the mean free path approaches a constant value. Thus, the temperature dependence of the thermal conductivity should mimic that of the specific heat. Since boundary scattering dominates at very low temperatures when phonon–phonon and phonon-defect scattering is weak, the thermal conductivity is expected to be of the form $\kappa \sim T^3$ and proportional to the sample size. This behavior is in fact obeyed in many single crystals and polycrystalline materials measured to low temperatures. A good example of the boundary-scattering effect for different sized crystals of bismuth [32] is shown in Fig. 6. The boundary scattering is particularly strong in this case because this semimetal possesses no isotopes, and the point-defect scattering can be made very small.

Casimir's simple temperature- and frequency-independent phonon-boundary scattering rate is valid only if the scattering is purely diffuse. What happens if this is not the case? It turns out that the nature of the scattering of phonons at the boundary depends on the size of the phonon wavelength relative to the average surface roughness. For "rough" surfaces the phonon wavelength is smaller than the size of the surface roughness, and indeed the scattering is diffuse. If the surface is "smooth," i.e., if the surface asperities are smaller than the dominant phonon

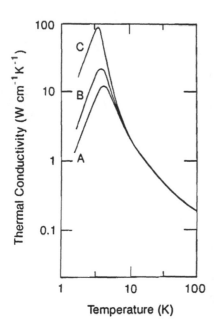

FIG. 6 Size effect of the thermal conductivity of bismuth single crystals of different cross sections, according to Ref. 32. Sample A, 1.2×3 mm^2; sample B: 2.8×3 mm^2; sample C: 8.8×8.6 mm^2.

wavelength, then at least some fraction of the phonons can be specularly scattered, and the resultant phonon-boundary scattering rate will be smaller than that given by Eq. (11). Experimentally this is manifested by a thermal conductivity falling more slowly than the specific heat, i.e., a mean free path slowly increasing above the Casimir value with decreasing temperature.

Berman [33] noted that for polycrystalline materials the boundary scattering effects could extend to quite high temperatures and depress the thermal conductivity by orders of magnitude; an example is shown in Fig. 7 for the case of Al_2O_3. Even though this effect obviously arises due to the limitation of the phonon mean free path to the relatively small values of the crystallite size (~1 μm as opposed to ~1 mm in the case of single crystals), it was generally found that the thermal conductivity increased more weakly than a T^3 law. Berman made the reasonable conjecture that the scattering at the crystallite boundaries was not perfect

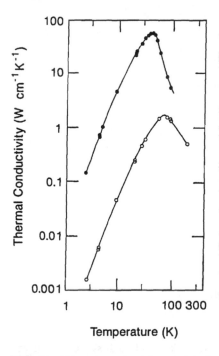

FIG. 7 Thermal conductivity of Al_2O_3. Upper curve, single crystal; lower curve, sintered polycrystal with grain size ~ 20 μm. After Ref. 33.

and that there was some transmission of phonons between crystallites that gave rise to the weakly increasing phonon mean free path with falling temperature.

This completes our brief discussion of the thermal conductivity of nonmetallic solids. To summarize, the thermal conductivity of a pure crystal will increase as T^3 at low temperature, pass through a maximum, and decrease exponentially or as $1/T$ at temperatures up to and above the Debye temperature. Both the magnitude of thermal conductivity at θ and the position in temperature of the peak increase with θ. If impurities are present, these produce characteristic temperature dependences or features in the thermal conductivity curve that depend on the nature of phonon-defect scattering.

In the next two sections we will specialize our discussion to the cases of the measured thermal conductivity of diamond single crystals and synthetically deposited films.

III. SINGLE-CRYSTAL DIAMOND

A. Introduction: The Classification of Diamonds

Natural diamonds are in fact quite pure, with the exception of contamination by nitrogen to levels as high as 0.3%; in rare cases, doping by boron can occur. Diamonds have long [34] been classified by type according to the nature of their infrared absorption spectrum. In a diamond crystal containing no defects, infrared absorption in the one-phonon region of the spectrum (<1330 cm^{-1}) is disallowed by symmetry. Defects destroy the local symmetry of the lattice and lead to the creation of dipoles, which give rise to absorption in this region of the infrared spectrum. The most common class of natural diamond is type Ia. In these diamonds nitrogen is present in fairly high concentrations in the form of pairs of atoms (A aggregates) or groups of several atoms (B aggregates). The characteristic infrared absorption of type IaA diamond exhibits peaks at 1197 and 1282 cm^{-1}; type IaB has peaks near 1175 and 1282 cm^{-1}. Most type Ia diamonds exhibit infrared spectra that are a linear combination of these, indicating the simultaneous presence of both A and B aggregates. Kaiser and Bond [35] and Lightowlers and Dean [36] both showed that the absorption at 1282 cm^{-1} is proportional to the nitrogen concentration; the exact contribution to this absorption from A and B aggregates was recently determined by Woods et al. [37]. They found for purely type

IaA diamonds that 17.5 ppm of nitrogen produces a 1 cm^{-1} absorption at 1282 cm^{-1}, whereas for type IaB diamonds 103.8 ppm nitrogen produces a 1 cm^{-1} absorption again at 1282 cm^{-1}. In addition to these features, some type Ia diamonds show absorption at 1370 cm^{-1}, which is attributed to "platelets." The question as to whether these platelets contain nitrogen has been the source of a great deal of controversy over the past 30 years. Today it is generally believed that the platelets contain little, if any, nitrogen and really are stacking faults. However, the 1370 cm^{-1} peak always occurs in conjunction with an A or a B feature suggesting that nitrogen may at least play a role in their formation.

Most nitrogen-containing synthetic diamonds but only about 1 in 1000 natural diamonds are classified as type Ib. In these stones the nitrogen is present in substitutional form and is characterized by a peak in the infrared absorption at 1130 cm^{-1}, the strength of which scales with the nitrogen concentration as determined by electron paramagnetic resonance [38]. It is also possible to have stones exhibiting mixed type Ia and type Ib behavior.

Semiconducting diamonds in which the active acceptor is thought to be substitutional boron are termed type IIb diamonds. These diamonds can have resistivities as low as 100 Ω cm and are characterized by infrared absorption at 1282 cm^{-1}, which increases strongly upon cooling to liquid nitrogen temperatures.

A diamond that shows no infrared absorption in the one-phonon regime is termed a type IIa diamond. This does not mean that this type of diamond is totally free of nitrogen or acceptor impurities, but rather that the impurity level is below that detectable using infrared spectroscopy. Type IIa diamonds have nitrogen concentrations below about 10^{18} atoms cm^{-3}. As we shall see, however, even nitrogen at this level is capable of influencing the thermal conductivity in a certain temperature range.

B. Type IIa Diamond: Overview

In 1911 Eucken [7] showed that the thermal conductivity of diamond was high compared to other dielectric solids. Serious study of the temperature dependence of the thermal conductivity did not occur, however, until well after Peierls's pioneering work on phonon scattering in solids, when the Oxford group, led by Berman, began investigating the thermal properties of nonmetallic materials, including diamond. The first measurements on type IIa diamond were reported by Berman et al. [18].

These measurements postdate an earlier set taken on Type Ia diamonds, which will be discussed later. Other measurements as a function of temperature on type IIa crystals have been reported by Berman et al. [39], shown in Fig. 8, and Berman and Martinez [40]. The results, on the whole, show the expected behavior of a typical dielectric crystal. Berman and Martinez attempted to fit their data to the Debye model, using the following forms for the scattering rates:

$$\tau_u^{-1}(x) = 1.31x^2T^4\exp(-270/T)$$

$$\tau_{pd}^{-1}(x) = 0.035x^4T^4 \tag{15}$$

$$\tau_b^{-1}(x) = 3 \times 10^6 \text{ sec}^{-1}$$

Note that the umklapp exponent corresponds to $b \simeq 7.5$, and the boundary rate is about two to three times smaller than that calculated using the Casimir formula. Berman and Martinez surmise that there may be substantial specular reflection of phonons at the boundaries that yields the lower than expected value; this will be discussed later. The point-defect

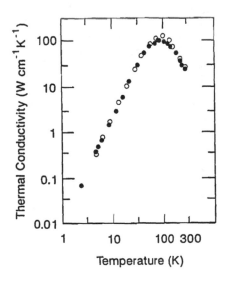

FIG. 8 Thermal conductivity of two natural diamond single crystals, according to Ref. 39.

term yields a scattering strength about ten times higher than that expected from scattering from ^{13}C isotopes at a naturally occurring level of 1.1%. The authors attributed this to additional point-defect scattering from nitrogen at a concentration on the order of 100 ppm. This is somewhat higher than the level of nitrogen normally found in type IIa stones (~20 ppm). Additionally, Berman and Martinez state that it was impossible for them to fit the data below the peak without including a scattering rate proportional to ω, which they attributed to the presence of dislocations.

Slack [41] measured two synthetic type IIa single-crystal diamonds made by the General Electric high-temperature high-pressure process; the data are reproduced in Fig. 9. In this case most of the nitrogen was present in dispersed form. In the sample designated R201, the nitrogen content was measured using paramagnetic resonance to be 1.5×10^{16} cm^{-3}, whereas that of sample R207 was 9×10^{18} cm^{-3}. Sample R201 had a peak value of 175 W cm^{-1} K^{-1} at 65 K, whereas that of R207 was

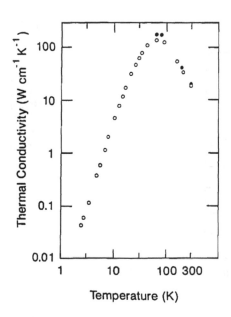

FIG. 9 Thermal conductivity of two synthetic diamond single crystals. Sample designation: (●), sample R201; (○), sample R207. After Ref. 41.

140 W cm^{-1} K^{-1} at 70 K. Both of these diamonds, as well as the type IIa diamonds measured by Berman, had room temperature values of 20 W cm^{-1} K^{-1}. It is not clear in the case of Slack's samples, in particular sample R201, whether the higher peak value of the thermal conductivity was due to a smaller point-defect scattering or smaller boundary scattering. Sample R201 had an average diameter of 3.8 mm, whereas the diameter of Berman's crystals was closer to 1 mm; this together with the fact that the peak occurs at lower temperature in Slack's samples (65 and 70 K) than in Berman's samples (~ 90 K) might suggest that indeed the boundary scattering was smaller in the synthetic crystals. If point-defect scattering were the dominant influencing factor at the peak, one would expect a depression in the value of the conductivity for higher nitrogen concentrations but no substantial shift in the peak position. Slack made no attempt, however, to analyze his results using the Debye model.

Burgemeister [42] measured a large series of natural type IIa stones at temperatures of 320 and 450 K. He found no statistically significant difference in conductivity amongst different samples, with the average values of the thermal conductivity being 19.3 ± 0.6 W cm^{-1} K^{-1} at 320 K and 12.5 ± 0.6 W cm^{-1} K^{-1} at 450 K.

The experimental data on the thermal conductivity of type IIa diamond were recently extended to higher temperature by Vandersande et al. [43]. Using a flash diffusivity technique, they measured the thermal conductivity between 500 and 1250 K of two natural stones obtained from DeBeers; the results are shown in Fig. 10.

Combining the data of Vandersande et al. above 500 K, those of Burgemeister between 320 and 450 K, and those of Berman below room temperature, one obtains the result shown in Fig. 11. Applying the Debye model to this composite curve is desirable since the data cover nearly three orders of magnitude in temperature and six orders of magnitude in conductivity. This procedure was recently carried out by Onn et al. [44] as part of their study of the thermal conductivity of natural abundance and isotopically pure diamond. Aside from the isotopic effect, which we will discuss in more detail later, they found that the high-temperature data could be fit with an umklapp term of the form

$$\tau_u^{-1} = 820x^2T^3\exp(-540/T) \tag{16}$$

and that a rate of the form used by Berman, namely $x^2T^4\exp(-b/T)$, could not fit the high-temperature data of Vandersande. The form used by Onn et al. is particularly appealing because it is the same form for the

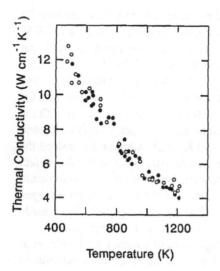

FIG. 10 Thermal conductivity of two type IIa diamond single crystals at high temperature. After Ref. 43.

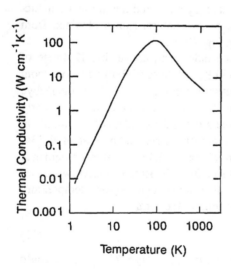

FIG. 11 Composite thermal conductivity curve of type IIa natural diamond between 1 and 1000 K.

umklapp rate used to fit the data of silicon and germanium [15], which are isostructural to diamond, and agrees with the form suggested by Herring [14] on theoretical grounds. In fact, from (16) one finds from the exponent $b \simeq 3.4$, which is nearly the same as that used for silicon and germanium. For their sample containing natural abundance of ^{13}C, Onn et al. also find a point-defect scattering rate coefficient of 0.04 compared with the value of 0.034 found by Berman and Martinez [40], both of which exceed the value of 0.0075 expected due to scattering by 1.1% ^{13}C.

It can be concluded that above about 10 K the measured thermal conductivity of type IIa natural diamond can be understood reasonably well within the framework of the Debye model. There are, however, some features that appear puzzling. First, the boundary rate used to fit the data at low temperature is about a factor of two smaller than that calculated using the Casimir formula. In the next section we will describe measurements made to lower temperatures that provide information about the nature of boundary scattering in diamond. Another area of interest is the extra point-defect scattering required to fit the experimental data. Even though there is no doubt that nitrogen is present even in type IIa stones, it is not clear from our discussion up to this point that this impurity in the known concentrations can account for the observed behavior, or if there is some other unidentified impurity that contributes to the point-defect scattering. Some light can be shed on this question by considering the thermal conductivity of more heavily nitrogenated samples, namely type Ia diamonds. This is the topic of Section III.F.

C. Type IIa Diamond: Boundary Scattering Effects

In 1938 deHaas and Biermasz [45] measured the thermal conductivity of two diamonds down to 3 K and reported that the thermal conductivity below 15 K fell as $T^{2.5}$. Although they suggested that this result might be due to the scattering of lattice waves off the boundaries of the crystal, no definite conclusions were drawn. Berman et al. [46] showed on measurements down to 2 K that the thermal conductivity did not indeed depend on sample size (an observation that by that time, had already been made on several other materials) but that the temperature dependence was only $T^{2.8}$, as opposed to the purely T^3 behavior Casimir predicted. In addition, the mean free path calculated from the thermal conductivity and the specific heat exceeded the sample cross-section and grew as the temperature was lowered. The authors argued that this was consistent with some fraction of the phonons undergoing specular reflection with the crystal boundaries and showed that at low temperatures the phonon wavelength did indeed exceed the extent of the surface roughness by a factor of ten.

A very careful study of the nature of boundary scattering in type IIa diamonds was carried out by Vandersande [47]. He studied five crystals with different surface finishes between 0.3 and 20 K. Unexpectedly, it was found that for polished surfaces the mean free path was nearly temperature independent but almost three times larger than the Casimir value. Upon roughening of the surface, the mean free path became temperature dependent but was on the order of the sample size. An example of one set of his data is shown in Fig. 12. Vandersande suggested that polished samples had a network of cracks just below the surface from which the phonons were scattered partially specularly but independent of frequency; roughening of the corresponding surface was then thought to heal the cracks and allow other phonons to interact directly with the boundaries. The existence of microcracks in polished diamonds had in fact been observed by Frank et al. [48].

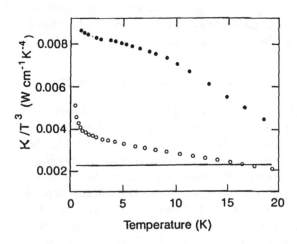

FIG. 12 κ/T^3 diamond IIa-1, from Ref. 47. Upper curve, smooth surface; lower curve, rough surface. Solid line represents the theoretical Casimir limit.

Another interesting experiment on type IIa diamonds in the boundary regime was performed by Vandersande [49]. He measured two diamonds with the direction of heat flow along the <100> and <110> axes, respectively, and observed that the conductivity along <100> was nearly twice that along <110>. The difference could not be accounted for by either specular reflection or scattering from strain fields associated with the polished surfaces; rather, the results were found to be in qualitative agreement with theoretical predictions of the focusing of phonon energy flow in the <100> direction in diamond [50].

D.   Type IIa Diamond: Scattering by Irradiation-Induced Defects

We have seen in much of the previous discussion how strongly defects can affect the thermal conductivity. On the other hand, determining exactly what defect may be doing the scattering can be a difficult chore, especially when very low concentrations (down to 1 ppm and even less) can still produce significant effects. One technique, which has met with some success, is to artificially introduce defects into a pristine crystal by irradiation with elementary particles and observe their effect on the heat conduction. This method has the advantage of providing a means of controlling the size, type, and concentration of the defects introduced. In addition, the information gained by studying the thermal conductivity of irradiated specimens is potentially important for applications in industry where the irradiation of these materials cannot be prevented.

Interestingly, whereas the optical properties of irradiated diamond have been studied in rather great detail, the thermal conductivity has remained relatively unexplored. Berman et al. [51] stated that the thermal conductivity of a diamond irradiated with γ-rays up to an unstated level remain unchanged. After irradiation to a level of 1.1×10^{18} neutrons cm^{-3}, however, the thermal conductivity was decreased by a factor of 50 at 90 K and 180 at 20 K. The authors made no conjecture as to the origin of the additional scattering.

A more thorough investigation was carried out above 320 K by Burgemeister and Ammerlaan [52] and Burgemeister et al. [53] for electron-irradiated diamonds. The latter study was particularly effective because it monitored the production of defects using optical techniques. If a diamond is irradiated with electrons of energy high enough to eject a carbon atom from the lattice but just not high enough to produce secondary displacements, the damage to a large extent is confined to the creation of vacancies and carbon interstitials. Even though there has been some evidence [54] that the interstitial carbon atom is mobile at and even below room temperature and that this could provide a means for interstitials to annihilate vacancies or agglomerate into larger defects, Burgemeister and Ammerlaan argued that at high temperatures these defects will be unimportant, most of the additional scattering being of the point-defect type due to vacancies. They observed an additional thermal resistance that scaled with the vacancy concentration calculated, assuming a displacement energy of carbon atoms of 80 eV. More substantial proof that vacancies indeed did play a strong role was provided by Burgemeister et al. [53]. They showed that the irradiation produced a neutral vacancy (called the *GR1 center*) that exhibits strong optical absorption at 1.673 eV; furthermore, the increase in resistivity of irradiated specimens above that of natural diamond scaled with the intensity of this absorption, which itself is proportional to the vacancy concentration; see Fig. 13. No attempt was made to analyze the results using the Klemens point-defect scattering formula, however.

Evidence that the interstitial carbon in electron-irradiated diamond does indeed agglomerate was presented by Vandersande [55], who studied three electron-irradiated diamonds between 0.3 and 20 K. In this temperature range the phonon wavelength is long enough to be fairly insensitive to point defects but quite sensitive to large-scale defects. An example of Vandersande's results is shown in Fig. 14. The strong rise in κ/T^3 at low temperatures was attributed to a crossover from geometrical to Rayleigh scattering characteristic of a precipitate or agglomerate. From the location in temperature of this rise and the strength of the scattering rate required to fit the data, Vandersande was able to determine the size and concentration of the aggregate scatterers. He found that for unannealed specimens, the aggregate size was on the order of 100 Å, but after a high-temperature anneal the aggregate size decreased to 37–55 Å. The total number of atoms comprising the agglomerates was in reasonable agreement with an estimate of the number of carbon atoms ejected from their lattice sites. Since aggregate scattering was observed even on a diamond irradiated at 77 K and subsequently annealed to room temperature, this was taken as evidence that the carbon interstitial is mobile at or below 300 K.

At the time that Ref. [55] was published, it was generally accepted that the displacement threshold for removing a carbon atom from the

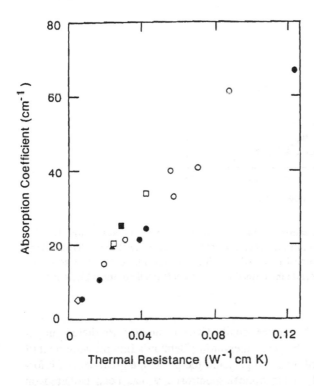

FIG. 13 GR1 absorption due to neutral vacancies versus thermal resistivity for various electron-irradiation energies. Open symbols are for samples irradiated below 250 K; closed symbols, for samples irradiated at 500 K. Symbols designate irradiating electron energy: circles, 1.50 MeV; triangles, 0.90 MeV; squares, 0.60 MeV; diamonds, 0.35 MeV. Adapted from Ref. 53.

diamond lattice was on the order of 80 eV; thus, the electron energy was adjusted to ensure that no secondary displacements occurred. Recent work [56] indicates, however, that this threshold energy is closer to 40 eV. Thus, there is the possibility that in Vandersande's experiment more than one carbon atom was released per incident electron, and this may have affected the results.

In order to gain a better understanding of the role of disordered carbon in the thermal conductivity of vapor-deposited films (a topic discussed in detail later), my colleagues and I [57] have recently carried out detailed

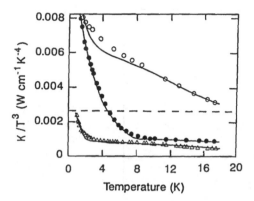

FIG. 14 κ/T^3 for electron-irradiated diamond at low temperature, from Ref. 55. Open circles are for unirradiated crystal; solid circles, after 12 h irradiation; open triangles, after an anneal at 1150°C. Solid lines are fits to the data assuming precipitate scattering, and the dashed line represents the theoretical Casimir limit.

measurements of the thermal conductivity of neutron-irradiated single-crystal diamond. In the years intervening Berman's first measurement of a neutron irradiated diamond [51] and the present, a great deal of information, principally using optical spectroscopy, has been gathered on the types of defects introduced into irradiated crystals. X-ray [58] and infrared absorption [59] studies indicate that low doses of neutrons ($<10^{18}$ cm^{-2}) produce vacancies and small regions of disordered carbon; for higher doses, amorphization develops. The creation of disordered regions for low doses occurs because each incoming neutron displaces at least several hundred carbon atoms, and these reform into regions of mixed double and single bonding [60]. The effect of irradiation on the thermal conductivity is shown in Fig. 15. We see that, contrary to the fairly subtle changes that occur in electron-irradiated material, the thermal conductivity of the neutron-irradiated material is drastically reduced and characterized by a strong dip in the curve in the neighborhood of 20–30 K. At the lowest temperatures, however, the curves seem to approach each other again. These results have been analyzed in terms of scattering of phonons from the precipitates of disordered carbon. The solid lines in Fig. 15 are fits to the data using the Debye model. The procedure used was to first fit the data of the unirradiated material using

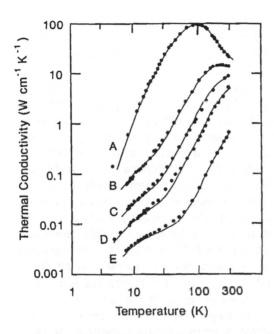

FIG. 15 Thermal conductivity of type IIa natural diamond irradiated with neutrons, from Ref. 57. Sample designation: A, pure crystal; B, 3×10^{16} cm^{-2}; C, 1.2×10^{17} cm^{-2}; D, 6×10^{17} cm^{-2}; E, 4.5×10^{18} cm^{-2}. Solid lines are fits to the data discussed in the text.

scattering rates similar to those used by Onn et al. [44] and described in Section III. Then the data of the irradiated material was fitted by adding a precipitate scattering term of the form of Eq. (13), leaving the other rates (except for the point-defect rate) unchanged from their single crystal values. We found that the concentration of the disordered regions was roughly equal to the neutron dose, i.e., each incoming neutron produced a disordered region. The size of the disordered region was estimated from the position of the dip in the $\kappa(T)$ curves to be approximately 15 Å and, thus, contains several hundred carbon atoms. This was in rough agreement with the total number of carbon atoms displaced per incoming neutron as estimated using simple radiation damage models. Further evidence in support of the formation of disordered regions was provided by the observation of irradiation-induced absorption in the one-phonon region of the infrared spectrum. From the strength of the absorption at 1200 cm^{-1}, a lower limit on the total number of displaced carbon atoms was derived, which also was in rough agreement with the thermal conductivity data.

It is of interest to discuss what led us to conclude that the dips observed in Fig. 15 in neutron-irradiated diamond were due to aggregate scattering and not to resonant scattering from vacancies. If the neutral vacancy in diamond did indeed scatter phonons resonantly, one ought to also observe this effect in electron-irradiated material, because in this case the vacancy is the primary induced defect. Unfortunately, in neither the Burgemeister nor the Vandersande study on electron-irradiated diamond already discussed did the authors measure κ in the region of the dip in neutron-irradiated diamond. Therefore, I recently obtained from Dr. Vandersande one of the diamonds used for his study. This sample had been irradiated to a total dose of 5×10^{16} electrons cm^{-2} and annealed at 1100°C. A luminescence spectrum revealed emission at 1.673 eV characteristic of the neutral GR1 vacancy. The thermal conductivity of this sample from 7–300 K is shown in Fig. 16. We see that although there is a reduction with respect to the single-crystal value, it appears to be due only to additional point-defect scattering (which diminishes the magnitude of the peak at 95 K) and a small amount of additional frequency-independent scattering at low temperature, most likely due to the aggregates found by Vandersande. Nowhere in this temperature regime is there an indication of a resonant-type scattering. We thus conclude that although there are GR1 vacancies present in *neutron*-irradiated material, they most likely act only as point-defect scatterers and do not cause the strong dip in the conductivity near 25 K.

E. Type IIa Diamond: Isotopic Effects

In 1942 Pomeranchuk [22] suggested that due to their mass difference isotopes would scatter phonons, and he calculated a mean free path for such a process. The isotope effect on the thermal conductivity remained unexplored for some time after that paper, however. Slack [21] analyzed experimental results on Si, Ge, and KCl and suggested that the thermal conductivity at the peak in these materials was limited by isotope scattering and that the effect may even be noticeable in diamond. Berman et al. [18] compared experimental results for nine different nonmetallic crystals with a variational calculation of the isotope resistance and found

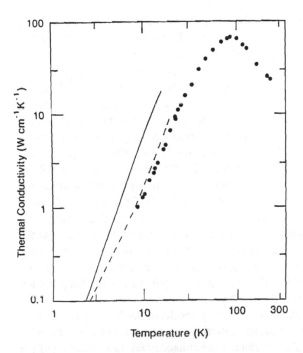

FIG. 16 Thermal conductivity of a natural type IIa diamond irradiated with electrons, as measured by the author. The sample was measured in the unirradiated (solid line) and irradiated (dashed line) states at low temperature, from Ref. 55.

reasonable agreement in all cases but diamond, for which it was necessary to invoke additional impurity scattering to fit the data. Since at that point in time the synthesis of diamond in the laboratory was just developing, it was not possible to vary the ^{13}C concentration systematically to study the effect in detail. Thus, it was left to Geballe and Hull [24] to provide the first concrete proof of the isotope effect in germanium.

In the years since these early investigations on isotopic effects in dielectric crystals, the synthesis of diamond in the laboratory improved slowly but surely, and by 1970 General Electric [61] used the high-pressure high-temperature belt process to produce gem-quality synthetic diamond single crystals that were large enough for thermal conductivity studies. These materials all contained the natural abundance of ^{13}C, however. It was not until the development of a new two-step process [62] incorporating chemically vapor-deposited diamond did General Electric undertake growing diamonds with reduced ^{13}C content.

Measurements of Anthony et al. [63] on a gem containing reduced amounts of ^{13}C indicated that a crystal containing 0.07% ^{13}C had a thermal diffusivity 50% higher than that of a diamond with a natural abundance of ^{13}C at room temperature. Assuming that the specific heat of the isotopically enriched sample is approximately the same as that of natural diamond, which was later verified by Morelli et al. [64], these authors derived a value of 33 W cm^{-1} K^{-1} for the thermal conductivity at 300 K. One way to estimate the influence of the additional point-defect scattering on the thermal conductivity is to use the additive resistivity approximation [65]. In this model the thermal resistivities arising from each scattering process are assumed to add in series. One finds that the added resistivity due to the presence of ^{13}C in the diamond lattice is only about 5% of the total diamond resistivity at room temperature. Anthony et al. thus concluded that their results were quite surprising but provided no explanation for the large enhancement.

An attempt to explain the diffusivity results of Anthony et al. based on a disorder-enhanced umklapp scattering mechanism was made by Bray and Anthony [66]. They attributed the enhancement to a smearing of the phonon dispersion by the local change in mass. Hass et al. [67], however, claimed that the Bray model was too crude and that the actual disorder-induced broadening was two orders of magnitude smaller. Hass et al. further argued that the observed enhancement could be explained using the standard Klemens formula Eq. (9), *provided sufficient normal scattering of phonon occurs*. They used the Callaway formalism to show that for appropriately chosen scattering rates the composition dependence of κ at room temperature could be explained, a conclusion that was reached independently by Nepsha et al. [68]. Berman [69] also showed that the results were consistent with the Callaway model in the limit of strong normal phonon scattering. The only drawback of these models is that one must choose a form for the normal scattering rate (as well as the umklapp rate) for which there is essentially no theoretical justification. Thus, this amounts essentially to fitting the data by introducing two more adjustable parameters (the n-process exponent and coefficient). For instance, Hass et al. use $\tau_n^{-1} = 0.011xT^5$ for reasons that are not clear.

Onn et al. [44] derived thermal conductivity data on diamonds of various isotopic composition from diffusivity measurements between

300 and 500 K; the results are shown in Fig. 17. In addition they reanalyzed existing data in light of the recent high-temperature measurements of Vandersande et al. [43]. As discussed in section III.B, they showed that the umklapp rate used by Berman to fit the data up to 300 K failed to fit the high-temperature data; Onn et al. suggested that the correct umklapp rate was of the form $\omega^2 T \exp(-\theta/bT)$, a form originally suggested by Graebner and Herb [70]. Using this form for the umklapp rate and setting the point-defect scattering rate to zero, they were able to obtain a very good fit to the data on the isotopically purified gem without invoking normal processes; see Fig. 17. However, the difference between the point-defect scattering rate coefficient for the isotopically purified and natural abundance synthetic diamonds was found to be five times greater than that due to ^{13}C mass difference scattering alone. Onn et al. thus suggested that the synthetic diamond containing ^{13}C impurities contained some other unidentified impurity or defect as well. Onn et al.

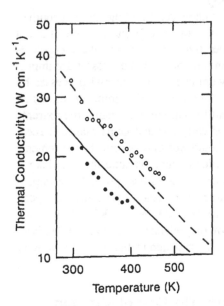

FIG. 17 Thermal conductivity between 300 and 500 K of isotopically pure (open circles) and natural abundance (closed circles) diamonds. The solid line is a fit to the Debye model assuming umklapp, boundary, and point-defect scattering. The dashed line is obtained by setting the point-defect rate equal to zero. Adapted from Ref. 44.

surmised that 70-ppm vacancies would be capable of producing the required additional point-defect scattering in the natural abundance sample by virtue of their mass difference alone and estimated that, if strain field effects are considered, as little as 1-ppm vacancies could provide the sufficient amount of scattering. It is not clear how this estimate of 1-ppm vacancies is arrived at. For a vacancy one would expect that the point-defect scattering parameter $(\Delta M/M + 2\Delta\zeta/\zeta) = -3$, as opposed to -1 if just mass difference scattering is considered [71]. Since this factor enters the scattering rate quadratically, it would lead to an estimation of $70/(-3)^2 \simeq 8$-ppm vacancies as providing enough additional point-defect scattering to explain the observed results. In any event, I would suggest that it might be useful to look very carefully at the optical absorption and emission as a function of ^{13}C concentration in the neighborhood of the GR1 neutral vacancy excitation at 1.673 eV to see if there is any signature indicating the presence of vacancies in samples containing the heavier carbon isotope.

Thus, it seems that although some enhancement of the thermal conductivity at room temperature is obtained by isotopic purification, it is not yet clear what mechanism is operative. Even though the increase is consistent with that expected when normal phonon processes are important, it could also be explained without invoking normal processes by requiring the presence of some other unidentified impurity or defect whenever ^{13}C is present in the lattice. On the other hand, the extreme crystalline perfection of the isotopically purified gems makes it seem unlikely that normal processes should *not* play a role. Measurements down to low temperature on isotopically pure diamond may shed some light on this situation, but these measurements may prove to be impossible due to the expected magnitude of the conductivity near the peak (predicted [72] to be up to or in excess of $1000\ W\ cm^{-1}K^{-1}$). At the time of this writing, the explanation of the enhanced thermal conductivity is currently an outstanding problem in diamond thermal conductivity research.

F. Type Ia and Ib Diamond: The Role of Nitrogen

The first detailed set of measurements on the thermal conductivity of diamonds was done on type Ia specimens, most likely because these occur in the greatest abundance in nature. Berman et al. [46] measured a type I stone that was successively ground and/or sawn to smaller dimensions to study the size effect. They showed that both the magnitude of the conductivity and the location in temperature of the maximum were higher than any other material measured up to that time and that for $T > 100$ K the behavior was in accord with that expected due to umklapp scattering. In the boundary regime they found that the mean free path exceeded the Casimir value and attributed the difference to specular reflection. For the present discussion the most salient feature of their investigation was the observation that the conductivity near the maximum (70–80 K) was not as high as predicted based on boundary and umklapp scattering alone and that near the peak some form of defect or impurity was contributing to the scattering. Based on the temperature variation of the observed conductivity, Klemens [73] suggested that the defects, unidentified at that time, occurred in groups of dimension on the order of 30–100 Å. In a later set of measurements on both type I and type II stones, Berman et al. [18] showed that isotope scattering was not large enough to explain the difference between the predicted and measured values, particularly in the case of the type I material, and concluded that some other impurity must be responsible for a large amount of scattering.

With the vacuum fusion analysis work of Kaiser and Bond in 1959 [35], it became clear that type I diamonds contained large amounts (up to 0.3%) of nitrogen. Slack [74] pointed out that the nitrogen in type Ia diamond is not electrically active so the scattering should be less at low temperature than in the case of Si and SiC, and this is indeed the case: Berman's results on type Ia diamonds were boundary limited below 10 K. Slack also stated that the nitrogen did appear to depress the conductivity in the range above 50 K but made no attempt to compare the existing data with theory because the exact nature of the nitrogen impurity state was not known.

To investigate more deeply the effect of nitrogen on κ, Slack [41] measured a natural type Ia crystal with a very large (3.5×10^{20} cm^{-3}) amount of nitrogen and found a room temperature value of 6.8 W cm^{-1} K^{-1} and a peak value of 19 W cm^{-1} K^{-1} at 85 K; see Fig. 18. Under the assumption that the additional thermal resistivity in a nitrogen-bearing diamond is proportional to nitrogen concentration, Slack showed that the minimum thermal conductivity for a diamond is 5.5 W cm^{-1} K^{-1}, based on a maximum solubility of nitrogen in diamond of 4.2×10^{20} cm^{-3}. Using the calculations of point-defect scattering by nitrogen in diamond by Agrawal and Verma [75], Slack indicated that mass

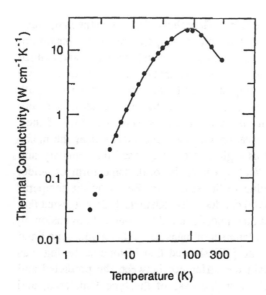

FIG. 18 Thermal conductivity of type Ia natural diamond single crystal containing 3.5×10^{20} cm^{-3} nitrogen. Data are from Ref. 41, and the solid line represents the theoretical model of Turk and Klemens from Ref. 76.

difference scattering alone could not account for the additional thermal resistivity, implying that scattering from the associated strain field must play the dominant role. The necessary degree of lattice mismatch required to fit the data, however, was about six times larger than that estimated from the work of Kaiser and Bond. Thus, Slack concluded that the nature of scattering by nitrogen in type Ia diamond was unclear.

Turk and Klemens [76] attacked the problem from a theoretical point of view by assuming that the nitrogen was divided between pointlike associations of a very few number of atoms and large disk-shaped platelets containing several thousand atoms. The existence of nitrogen platelets had been postulated based on transmission electron microscopy observations [77] of planar faults. They derived the following formula for the phonon–platelet scattering rate:

$$(\tau_{plat})^{-1} = (8H/v)\,\pi R^2 H/V_c(\Delta M/2M + \gamma\alpha)^2(k_B/h)^4 T^2 x^2 I(z) \qquad (17)$$

where R and H are the platelet radius and thickness, respectively; V_c, the platelet volume; and α, the fractional volume difference between nitrogen and carbon atoms. The function $I(z)$ is defined as

$$I(z) \simeq 1 - 0.5[J_0^2(z) + J_1^2(z)] - (1/z)J_0(z)J_1(z) \qquad (18)$$

with J_0 and J_1 the zeroeth- and first-order cylindrical Bessel functions and $z=Rk_BT/hv$. Turk and Klemens were able to fit the data on Slack's type Ia crystal (see Fig. 18) by assuming that 97% of the nitrogen was in pointlike associations containing eight nitrogen atoms, with the remainder in platelets one atom thick and 100 Å in diameter, each platelet thus containing approximately 5000 nitrogen atoms.

Berman et al. [39] measured seven type Ia diamonds and analyzed their results in terms of various scattering processes. They concluded, in concurrence with Turk and Klemens, that most of the impurity scattering was due to point scattering from groups of on the order of ten nitrogen atoms. On the other hand, by comparing the nitrogen concentration derived from a fit to the thermal conductivity with that determined from infrared absorption, Berman et al. concluded that the platelets contained little if any nitrogen and, in any case, provided very little scattering. Berman and Martinez [40] suggested that the "platelets" were in fact planar defects consisting of sheets of carbon atoms, i.e., stacking faults. Correlation of x-ray spike intensities with the total platelet area [78] forms the basis for the present opinion that in fact the platelets do indeed contain little, if any, nitrogen.

Over the past decade or so, ever-refined optical investigations [79] have established that the most common form of nitrogen impurity in type Ia diamond (and indeed in type IIa diamond at much lower concentrations) is the A aggregate consisting of two substitutional nitrogen atoms. Thus, it would seem that the analyses used by Turk and Klemens [76] and Berman et al. [39] overestimate the number of atoms in a pointlike cluster by a factor of four to five. Upon irradiation at least some fraction of the A aggregates are transformed into the H3 center, which consists of two nitrogen atoms surrounding a vacancy. However, it is well known [80] that even unirradiated type Ia and type IIa diamonds exhibit the H3, indicating that this defect may occur at fairly large concentrations naturally (the diamond studied by Morelli et al. [57] in fact showed H3 luminescence prior to irradiation; see Fig. 19). Thus, I suggest that in modeling the thermal conductivity one ought to also include a scattering rate due to H3 aggregates. Even in fairly small concentrations, these may

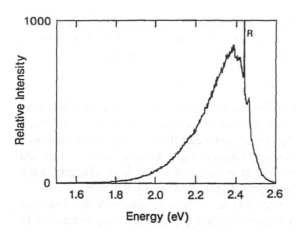

FIG. 19 H3 luminescence in a diamond characterized as type IIa according to its infrared spectrum. The line marked R is the zone-center Raman line. The H3 feature arises from two nitrogen atoms bound to a vacancy.

provide substantial scattering due to the large strain effect arising from the broken bonds surrounding the vacancy.

Another study along the lines of Berman et al. [39] was carried out above 320 K by Burgemeister [42]. He studied 49 type Ia diamonds and was able to correlate the value of the thermal conductivity at 320 K with the temperature exponent determined between 320 and 450 K. He additionally showed that the value at 320 K scaled with the infrared absorption peaks at 1282 and 1205 cm^{-1}, as well as the integrated absorption between 1000 and 1430 cm^{-1}. Using the relation between absorption at 1282 cm^{-1} and the concentration of A and B aggregate nitrogen, Burgemeister further showed a correlation between the conductivity at 320 K and the total nitrogen concentration $N_A + N_B$ (see Fig. 20) and concluded that in this type of diamond nearly all the impurity scattering is due to nitrogen in either the A or B forms.

In hopes of shedding light on the question of nitrogen aggregate size, Vandersande [81] measured two type Ia diamonds from 1 to 20 K. One diamond was classified as a IaA diamond and also exhibited an infrared absorption feature at 1370 cm^{-1} attributed to platelets. A second diamond contained only B features in the infrared and was thus classified as type IaB. Vandersande fitted the data shown in Fig. 21 to the Debye model

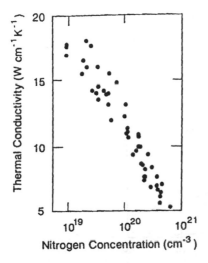

FIG. 20 Thermal conductivity of 49 type Ia diamonds at 320 K versus nitrogen concentration as determined from infrared absorption features, according to Ref. 42.

and showed that for the IaA diamond a large point-defect scattering rate as well as a rate corresponding to scattering from stacking faults were required. The first of these rates was consistent with the A feature being due to very small pointlike aggregates of a few nitrogen atoms, and the second rate was consistent with the observation of platelets. For the IaB diamond, however, Vandersande found evidence of scattering by aggregates of dimension 44–65 Å and concentration $(0.75–1.7) \times 10^{14}$ cm^{-3}. This size of aggregate was consistent with electron microscope observations [82] on type IaB diamonds, which suggested that aggregates of nitrogen on the order of 50–80 Å are responsible for the B feature, a hypothesis that was in complete contrast to the more widely held belief that the B feature was due to groups of only a few nitrogen atoms. Berman and Martinez [40] also suggested the existence of aggregates on the order of 60 Å based on the observation of a slight upturn in the thermal conductivity of a type Ia diamond at 10 K.

Although the effect of nitrogen on the thermal conductivity of type Ia diamond is complicated by the presence of platelets and the possibility of two different arrangements of nitrogen atoms (A and B features), the

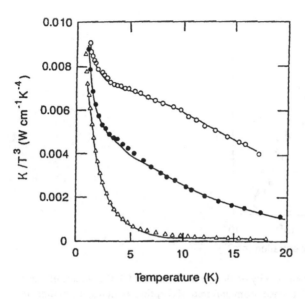

FIG. 21 κ/T^3 versus temperature (from Ref. 81) for a type IIa diamond (O), a type IaA diamond (●), and a type IaB diamond (Δ). Solid lines are the fits discussed in the text.

case of type Ib diamond should be more straightforward since the nitrogen is supposed to be present as a single substitutional impurity, although many crystals classified as type Ib also contain A and B aggregates. Berman et al. [39] studied 5 type Ib diamonds whose infrared absorption spectra were also measured. Since a single substitutional nitrogen atom is paramagnetic, its concentration can be determined by spin resonance measurements. Berman et al. found that the number of nitrogen atoms required to fit the thermal conductivity data (2×10^{20} cm^{-3}) exceeded both the concentration of paramagnetic nitrogen (10^{17}– 10^{18} cm^{-3}) and total nitrogen concentration (estimated from the absorption at 1282 cm^{-1} as $< 3 \times 10^{19}$ cm^{-3}). They pointed out that Sellschop et al. [83] found appreciable amounts of nickel (~1 ppm) in two type Ib diamonds and suggested that, due to the large mass difference between nickel and carbon, this impurity could produce the same scattering as about 100-ppm nitrogen and would be expected to scatter even more if the distortion of the lattice by the nickel atom is taken into account.

In his study of the thermal conductivity of type IIa and type Ia diamonds, Burgemeister [42] also measured the thermal conductivity of three type Ib diamonds. He found that the measured conductivity and the total concentration of nitrogen as estimated from the strengths of the 1282 and 1130-cm^{-1} features in the infrared were consistent with the type Ia results shown in Fig. 20. Nepsha et al. [84] indicated, however, based on a point-defect scattering rate formula similar to Eq. (10), that scattering by substitutional nitrogen would not be strong enough to account for Burgemeister's observations and that the conductivity must be limited by other types of defects.

To summarize this section, there is little question that nitrogen in natural diamonds influences the conductivity from perhaps above about 10 K to well above room temperature. Even though the effect can be understood at least qualitatively based on point-defect scattering, quantitative agreement between the concentration and distribution of nitrogen as calculated from fitting the thermal conductivity and that obtained by other means has not been achieved; in particular, there is a large discrepancy in the number of nitrogen atoms comprising the B aggregate as determined from optical and thermal conductivity studies. The principle problem in this author's opinion seems to be the simultaneous occurrence of nitrogen in several different forms (A and B aggregates and H3 centers), as well as the presence of other impurities and defects, principally transition metals and "platelets," or stacking faults. This problem might be overcome by a careful study of the conductivity of diamonds that show features in the optical spectra associated with only a single type of nitrogen-related defect and without the added complication of platelets; the specimens used by Woods et al. [37] for their infrared study are an example. Large specimens with these features are rare in nature, however, and probably cannot be produced in a controlled manner synthetically.

G. Type IIb Diamond

Type IIb diamonds are semiconducting with the acceptor species thought to be boron. In terms of thermal conductivity, one might anticipate the possibility of a small amount of phonon-carrier scattering to be present. It should be noted that even in type IIb diamonds of the highest electrical conductivity ($\sigma \sim 0.01$ Ω^{-1} cm^{-1}), the *electronic* contribution to the thermal conductivity κ_e is still negligible in comparison to the phonon part. This can be seen by applying the Wiedemann–Franz law to obtain an estimate of the upper limit to κ_e at room temperature:

$$\kappa_e = L_0 T_\sigma \tag{19}$$

where $L_0 = 2.45 \times 10^{-8} \text{ W } \Omega \text{ K}^{-2}$. This yields $\kappa_e < 10^{-3} \text{ W cm}^{-1} \text{ K}^{-1}$.

Much less is known about the thermal conductivity of type IIb diamond than other types of diamond. Berman et al. [18] measured a sample down to 3 K and found that from 100 K to room temperature this sample had a value only about two-thirds that of a type IIa specimen. At low temperatures the phonon mean free path was found to be less than the Casimir value, and the authors suggested that the difference could be due to scattering of phonons by carriers. This possibility was not pursued any further.

As part of his comprehensive study of the thermal conductivity of diamonds in the boundary regime, Vandersande [85] measured three type IIb diamonds from 0.3 to 20 K. As in the case of type IIa diamonds, he found significant specular reflection of phonons; in addition, however, Vandersande reported the existence of two dips in the conductivity curve, one at 5.5 K and the other in the neighborhood of 1 K; see Fig. 22. It was tentatively suggested that this might be due to a resonant

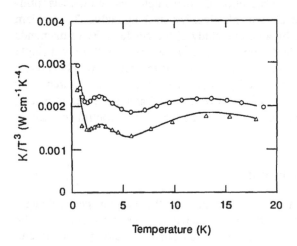

FIG. 22 κ/T^3 versus temperature for two type IIb diamonds. The solid lines are fits assuming a combination of boundary and resonant scattering. Adapted from Ref. 85.

interaction between the phonons and the acceptor, although no further evidence to substantiate this claim was provided.

Burgemeister [42] measured a set of 18 type IIb diamonds from 320 to 450 K and found very little variation in the thermal conductivity between the samples, even though the electrical resistivity of the samples spanned five orders of magnitude. At 320 K the average thermal conductivity was approximately 19.3 W cm^{-1} K^{-1} at 320 K and 12.5 W cm^{-1} K^{-1} at 450 K, i.e., the same as that of type IIa bars within the experimental error. Burgemeister concluded that there was negligible point-defect scattering due to the presence of boron or aluminum acceptors and placed upper limits of 200 and 1 ppm, respectively, for their concentrations.

IV. CHEMICALLY VAPOR-DEPOSITED DIAMOND FILMS

A. Introduction and Survey of Results Near Room Temperature

The synthesis of diamond in the laboratory at General Electric by Bundy et al. [86] was done at high pressures and temperatures where diamond is the stable form of carbon. Growth in the metastable thermodynamic state at low pressures and temperatures was in fact carried out by Eversole at Union Carbide Corporation *before* General Electric's famous high-temperature process was announced [1]. A review of the growth of diamond in the metastable state was recently given by Angus and Hayman [1]. Briefly, the roots of the present chemical vapor deposition (CVD) diamond process can be traced back to the seminal work of Deryagin and co-workers in the former Soviet Union, who used a variety of techniques including the presently favored method of growth from hydrogen/hydrocarbon mixtures from the vapor phase.

Because the thermal conductivity of diamond films depends so sensitively on defect and grain structure, which in turn is a strong function of the growth process, we will briefly outline the three basic types of CVD diamond growth techniques that are most widely used today. In *hot-filament chemical vapor deposition* (HFCVD) the hydrocarbon/hydrogen mixture is flowed into the growth chamber and over a filament (tantalum or tungsten are popular materials) heated to on the order of 2000–3000 K. Dissociation of the gases ensues with carbon deposited on a substrate, most commonly silicon or molybdenum, held at a specified temperature. The apparatus is low cost but has the disadvantage of introducing metal contaminants from the filament into the deposited film. In plasma-assisted chemical vapor deposition (PACVD), the hydrocarbon mixture is cracked using a high-frequency plasma discharge, eliminating the problem of filament impurity incorporation; this apparatus is generally more costly, however. A third growth technique is known as *DC plasma chemical vapor deposition* (DCPCVD). In this case, a DC arc between the gas distributor and substrate holder achieves the dissociation of hydrocarbon species. Growth rates on the order of 20 μm h^{-1} are achieved with this process, and DC plasma jet techniques [87] claim growth rates as high as 80 μm h^{-1}.

The initial attempts to grow CVD diamond films proceeded at very low growth rates and produced films too thin for conventional thermal conductivity studies. The first measurement of the thermal conductivity of a diamond film was carried out by Ono et al. [88] and Nishikawa et al. [89] using an infrared (IR) thermography technique between 100 and 130°C. They measured eight PACVD films ranging in thickness from 7 to 30 μm. In addition, the methane concentration in the growth gas was varied from 0.1 to 3.0 vol%. The films had been removed from the silicon substrate by dissolving away the latter in fluoric acid. Ono et al. found a rapid increase in thermal conductivity as the methane concentration was reduced below 1%; see Fig. 23. The highest conductivity obtained was 10 W cm^{-1} K^{-1} for a film grown at 0.1% methane concentration; the authors do not state to what temperature this value corresponds. In an attempt to correlate the thermal conductivity with the microstructure, Ono et al. obtained Raman spectra for each of the films studied. For the film grown at a low methane concentration they observed a strong line at 1333 cm^{-1}, which is the well-known zone center diamond phonon characteristic of sp^3-bonded carbon. For higher methane concentrations the 1333 cm^{-1} lines slowly disappeared, and a new broad feature centered at 1500 cm^{-1} appeared. This line is characteristic of a "graphitic," or sp^2-bonded carbon arrangement. Ono et al. suggest that the decrease in thermal conductivity may be caused by microscopic disorder as characterized by the ratio of the sp^2 and sp^3 Raman peak intensities.

Since the paper of Ono et al. was published in 1986, several other measurements of the thermal conductivity or thermal diffusivity (from which the conductivity is inferred) have been carried out at or near room temperature. Sawabe and Inuzuka [90] reported the thermal conductivity

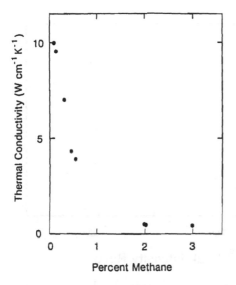

FIG. 23 Thermal conductivity as a function of methane concentration in the CH_4/H_2 growth gas for CVD diamond films, according to Ono, Ref. 88.

at 300 K of a film grown by a modified HFCVD process. They found a value of 11 W cm^{-1} K^{-1} and attributed the difference between this value and that of type IIa diamond as being due to the presence of amorphous carbon.

Herb et al. [91] used a steady-state technique employing radiation thermometry in the temperature range 340–390 K to measure a series of diamond films grown by both PACVD and DCPCVD. For both of these types of films, they found a correlation between the thermal conductivity and the sp^3/sp^2 Raman peak heights (Fig. 24). They also found a correlation between the full width at half maximum (FWHM) of the diamond Raman line and κ. For single-crystal type IIa diamond, Herb et al. measured a FWHM of 3.3 cm^{-1} at 100°C, whereas the values for their films were in the range 4.6–12.2 cm^{-1}. These results are shown in Fig. 25.

Using the thermal mirage technique, Pryor et al. [92] also found a correlation of the thermal diffusivity of four films and their graphitic content as estimated from the sp^2/sp^3 peak ratio method. Albin et al. [93]

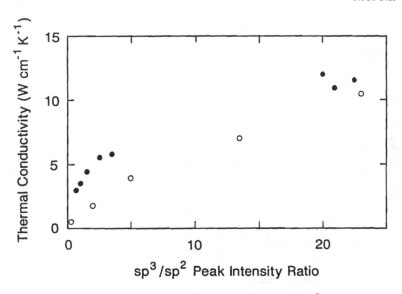

FIG. 24 Dependence of thermal conductivity at 100°C on sp^2-carbon content in a diamond film as determined from Raman spectra. Closed circles, data of Ono et al. [88]; open circles, data from Ref. 91.

performed diffusivity measurements on two PACVD films and derived conductivity values in the range 13–15 W cm^{-1} K^{-1}. They indicated the presence of some nondiamond sp^2-bonded carbon in their samples but drew no conclusions as to its influence on the thermal conductivity.

A similar dependence of thermal conductivity on methane concentration was found by Baba et al. [94] as was reported in the study by Ono, but the conclusions of Baba et al. for the dependence of methane concentration were different. These authors found an exponential increase in hydrogen content in the films (as determined by Rutherford backscattering) with increasing CH$_4$ content and observed peaks in the infrared spectrum, which they attributed to C–H stretch bands. They suggested that the impurity hydrogen disrupted the C–C bonds in the diamond lattice and thereby reduced the thermal conductivity.

Although the studies previously reported are important because they can provide a fairly simple determination of the thermal conductivity or diffusivity of a film near room temperature, they are clearly of limited value in investigating the defect structure of diamond films. As we have

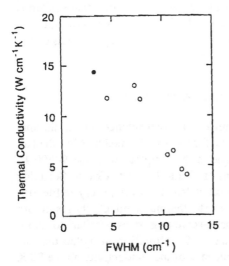

FIG. 25 Thermal conductivity at 100°C of CVD diamond as a function of the diamond Raman line full width at half maximum. Solid symbol is for single crystal. After Ref. 91.

seen already in the case of single-crystal diamond, minute amounts of impurity can strongly affect the thermal conductivity over a wide range of temperatures both above and below 300 K. In the case of diamond films, even the best specimens grown to date contain many different impurities and defects; these include not only nondiamond carbon, but also filament impurities, vacancies, carbon interstitials, hydrogen, nitrogen, and dislocations. It is somewhat risky to attribute a depression in the thermal conductivity due to the influence of a particular defect based on data at only one temperature or over a very limited temperature range. As in the case of single crystals, the most information can be gained only by measuring the conductivity over a wide temperature range (on a logarithmic scale) and studying the resulting temperature dependence. In fact, one of the most important "defects" in CVD diamond films from the viewpoint of heat conduction is their polycrystallinity, a feature that, for the most part, seems to have been overlooked in the majority of the studies previously mentioned. In the next two sections we will describe thermal conductivity measurements made as a function of temperature down to low temperatures and analyze the results in terms of phonon-scattering processes. We will divide the discussion into two parts: effects due to impurities and defects and anisotropic effects due to boundary scattering.

B.   The Effect of Impurities and Defects

The first measurement of the temperature dependence on the thermal conductivity of a synthetic diamond film was performed by Morelli et al. [95] using a steady-state technique in the temperature range 9–300 K. They measured two HFCVD films of 13 and 18 μm thickness. Scanning electron microscopy studies indicated that the average crystallite size was on the order of a few microns. The thermal conductivity results are shown in Fig. 26. Using phonon-scattering rates similar to those employed by Berman and Martinez [40], Morelli et al. noted that the magnitude and temperature dependence of the conductivity above 30 K was consistent with a predominance of phonon scattering by crystallite boundaries. The persistence of boundary scattering at room temperature and even higher had been predicted a few years prior to this study by Nepsha et al. [96]. Below 30 K, however, Morelli et al. found that the thermal conductivity deviated from the T^3 law expected for strictly diffuse scattering at a boundary. This effect was attributed to a small amount of disordered carbon in the lattice as determined from an sp^2-bonded carbon peak in the Raman spectrum. Finally, it was noted that the diamond films measured in this study had a thermal conductivity that exceeded that of prototypical high thermal conductivity materials such as copper, gold, silver, platinum, aluminum, and sapphire, even down to 150 K.

Nepsha et al. [97] proposed that the weakening of the temperature dependence of the thermal conductivity below 30 K found in the previous study could be explained by either specular reflection at crystallite boundaries or scattering from inclusions of nondiamond carbon in the lattice. In the case of specular reflection, the thermal conductivity was predicted to maintain the weak temperature dependence to temperatures below 10 K, whereas for the case of inclusion, or precipitate, scattering the thermal conductivity would revert to a T^3 law at lower temperatures. Nepsha et al. estimate that these regions would have to be about 12 Å in extent and have a concentration on the order of 1×10^{18} cm^{-3}. The data of Morelli et al. were taken down to 9 K, and there is just a hint that a

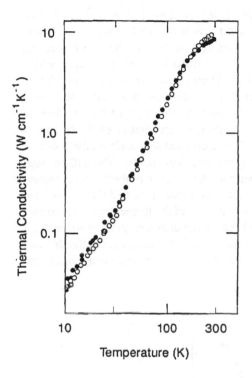

FIG. 26 Thermal conductivity of two HFCVD films 13 (●) and 18 (○) μm thick, according to Ref. 95.

faster temperature dependence may, in fact, be returning at low temperature, but the data do not extend to a low enough temperature to show this unequivocally.

Anthony et al. [98] measured the thermal conductivity of a HFCVD film 170 μm thick and containing a reduced amount of ^{13}C isotope in hopes of observing enhancement in κ due to the isotopic purification. The growth rate was only 0.32 μm h^{-1} due to the high cost of the isotopically pure starting materials. As a result the tungsten filament contamination of the film was at a fairly high level of 10^{19} atoms cm^{-3}. They noted a columnar structure to the grains, with the grain size at the substrate surface on the order of a few microns and that at the growth surface on the order of 75 μm. From etch pit studies, they deduced a dislocation density within the grains of greater than 10^8 cm^{-2}. The thermal conductivity was measured between 6 and 100 K using a steady-state technique that determined the conductivity in the film plane, and between 80 and 300 K using an AC diffusive wave technique, which measured κ across the film plane. Their results are shown in Fig. 27. In the overlap region Anthony et al. found that the measured conductivity was lower across the film than along it. This was attributed either to experimental error due to the combined uncertainties of the two techniques or to the fact that the AC measurement was made on the substrate side of the sample where the grain size was smaller. The authors suggested that the observed temperature dependence of the thermal conductivity of their films may be due to the dislocations and their associated strain fields. In this model the weakening of the temperature dependence of κ below about 30 K occurs because the dominant phonon wavelength begins to exceed the extent of the strain fields surrounding the dislocations. Anthony et al. finally concluded that any enhancement in thermal

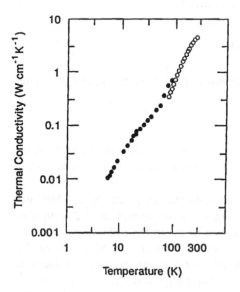

FIG. 27 Thermal conductivity of an HFCVD diamond film grown with isotopically pure methane, according to Ref. 98. Open circles, κ measured perpendicular to the film plane; closed circles, κ measured parallel to the film plane.

conductivity due to isotopic purification in their films is overshadowed by the large phonon-defect scattering.

In spite of some difference in absolute magnitude, the results of Anthony et al. were very similar to those of Morelli et al. in that the former also observed a strong weakening in the temperature dependence of κ below 30 K. The measurements of Anthony et al. extended to a slightly lower temperature, however, and did in fact indicate a return to a stronger temperature dependence at lower temperatures. It thus appears that the weak temperature dependence near 30 K is due to a "dip" in the thermal conductivity in the temperature range of 10–50 K. Unequivocal evidence for the occurrence of a dip in the thermal conductivity of diamond films at low temperatures was provided by Morelli et al. [99]. They measured six different films, three grown by HFCVD and three grown by PACVD; the films ranged in thickness from 250 to 600 µm. Raman spectroscopy on these films revealed only a diamond peak at 1332 cm^{-1} with no evidence of disordered or graphitic carbon. The authors also used infrared spectroscopy to characterize these films. For PACVD films they found evidence of only sp^3-bonded carbon with little or no evidence of C–H stretch vibrations or single phonon absorption indicating the presence of defects. The HFCVD films, however, exhibited weak absorption in the one-phonon portion of the IR spectrum, which the authors attributed to a small (< 1000 ppm) fraction of disordered carbon material. The presence of a very small amount of non-diamond carbon was indicated additionally by the dark appearance of the HFCVD films; two of the PACVD films, which showed no one-phonon absorption, appeared colorless, whereas a third PACVD film, which contained some sp^2 carbon, was dark. Morelli et al. also stated that secondary ion mass spectroscopy indicated the presence of tantalum (the filament material) in all three HFCVD films at a level of 50–100 ppm. The thermal conductivity results are shown in Fig. 28. There is a striking difference in the magnitude and temperature dependence of the thermal conductivity of the HFCVD samples and the two colorless PACVD samples. The HFCVD films had room temperature conductivities in the range 12–14 W cm^{-1} K^{-1} and a peak in κ near 250 K; additionally, they possessed a strong dip in the conductivity near 50 K. The two colorless PACVD films, on the other hand, had room temperature conductivities of 17 W cm^{-1} K^{-1} and a peak near 150 K; no evidence of a dip in κ occurred in these samples at low T. A third PACVD sample containing a small amount of sp^2-bonded carbon had a lower room temperature

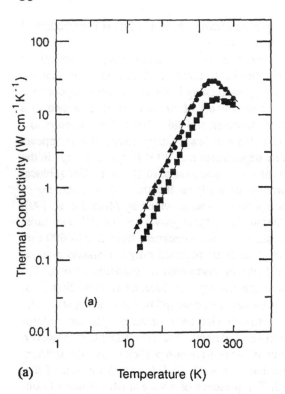

(a)   Temperature (K)

FIG. 28 Thermal conductivity of high-quality thick diamond films, from Ref. 99. (a) Films grown by PACVD and exhibiting no one-phonon infrared absorption; (b) films grown by HFCVD and exhibiting infrared absorption in the one-phonon region.

conductivity (14 W cm^{-1} K^{-1}) and higher peak temperature (200 K) but also showed no low-temperature feature. Morelli et al. suggested that the strong "dip" in the conductivity of the three HFCVD films was due to either resonant scattering from a defect or precipitate scattering from an extended defect (at the time of writing ref. [99] Morelli et al. were unaware of the model of inclusion scattering put forth by Nepsha et al. and described previously). It was shown that either model could fit the data and that it was impossible to distinguish which type of scattering was taking place based only on the thermal conductivity data. Morelli

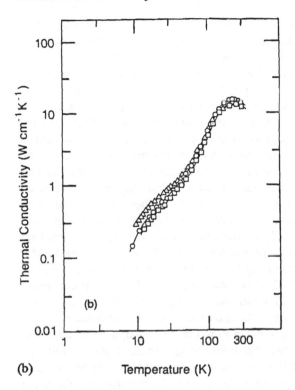

(b)   Temperature (K)

et al. suggested that the dip might be due to resonant scattering from a vacancy or from low energy modes associated with sp^2-bonded carbon or that regions of sp^2-bonded material could act as precipitate scatterers. From a fit to the thermal conductivity data, these disordered regions would have to be on the order of 5–10 Å in diameter. Because the dip occurred only in HFCVD films, scattering from precipitates of tantalum were suggested as a possibility as well.

The difficulty that Morelli et al. encountered in this study was that, although there was clear evidence of an additional strong phonon scattering due to defects in the HFCVD films, the complex defect structure of these films (even though they represented state-of-the-art diamond) made it almost impossible to attribute the scattering to a single defect. In addition, the crystallite size can differ from film to film and even across the thickness of a single film, making the selection of the proper boundary scattering rate a difficult one. It was these complications that led my colleagues and myself to study the thermal conductivity of neutron-irradiated diamond. As discussed in Section III.C, neutron irradiation provides a means of introducing into a single crystal some of the important defects that occur in CVD diamond. Since in a single crystal the boundary scattering rate is well defined and does not change with irradiation level, one can study the influence of these defects on the thermal conductivity in a much more controlled manner in contrast to measuring different CVD films with different defect morphologies. The successful modeling of the thermal conductivity of neutron-irradiated material in terms of precipitate scattering from regions of disordered carbon (Fig. 15) and the similarity with the thermal conductivity of HFCVD films would seem to suggest that precipitates of nondiamond carbon are the source of the additional phonon scattering and resultant dip in the thermal conductivity of these films. Additional support for this model is found by examining the infrared spectra of the films. The IR spectrum of the PACVD films containing no sp^2-bonded carbon appear nearly the same as type IIa diamond, whereas for the HFCVD diamond a continuum of absorption in the one-phonon region, which bears a striking resemblance to that found in an irradiated crystal, is observed. However, it must be stressed that even though the most likely additional scatterer in these HFCVD films seems to be disordered carbon, this is by no means a definite conclusion. Although we have tentatively eliminated resonant scattering from vacancies as being the source of this dip, further study of this interesting possibility or other unidentified defects is still required. Another possibility is that the similarity between HFCVD films and neutron-irradiated diamond thermal conductivity is only coincidental and that other mechanisms, such as resonant or precipitate scattering from tantalum, are active in the CVD material. An argument against the influence of filament impurities is provided by a measurement of Graebner and Herb [70] on a PACVD film. Even though this film does not contain any filament materials, it possesses a small but visible dip in conductivity near 30 K. I have also observed the occurrence of a low-temperature dip in PACVD films, but it seems to occur only on films in which there is a nonzero sp^2-carbon content based on one-phonon IR absorption. The question of the influence on the conductivity of small regions of sp^2-bonded carbon in an sp^3-bonded lattice is an interesting one that deserves further attention from both theoretical and experimental points of view.

Implicit in the preceding discussion of the results of Morelli et al. on the thermal conductivity of CVD diamond and neutron-irradiated diamond is that *a correlation between the strength of the sp²-bonded carbon Raman feature and the conductivity is not in general possible.* This is evident by noting that, in the case of the films, the thermal conductivity amongst the films differs by 30% at room temperature and an order of magnitude at 30 K, whereas the Raman spectrum shows no sp^2 features. Additionally, the neutron-irradiated diamond data show that the room temperature conductivity can be as low as a few W cm^{-1} K^{-1}, whereas the Raman spectra looks no different than that of a pristine type IIa diamond. This destroys any hopes of a general correlation between the ratio of Raman peak height intensities and the thermal conductivity. The fact is that, due to its high Debye temperature, the thermal conductivity of diamond is influenced more than any other material by defects at room temperature, and only a few parts per million of certain types of defects and impurities can drastically alter the conductivity without influencing at all other measured properties like the Raman spectrum or x-ray diffraction pattern.

If there is in general no correlation between the thermal conductivity of a film and the ratio of sp^2- to sp^3-bonded carbon peak heights, perhaps a correlation between the conductivity and the FWHM of the diamond Raman line, as has also been suggested [91], can be retained. Figure 29 shows the dependence on the thermal conductivity of a neutron-irradiated single-crystal diamond over that of the unirradiated material as a function of the change in the Raman line width [57]. We see that in this case, although a definite correlation does exist, it is quite different than that found by Herb et al. [91] shown in Fig. 25. One should not in general expect a correlation between the thermal conductivity of a film and the FWHM of the diamond Raman line. The width of the diamond Raman line is determined by the amount of "disorder" in the lattice. This disorder can come from the presence of defects, impurities, and even crystalline grain boundaries. The measured FWHM in a film is the sum of the contributions of each type of disorder to this width. For the thermal conductivity, each type of disorder produces some additional thermal resistivity, and the measured thermal conductivity is approximately the inverse of the sum of these individuals resistivities and the intrinsic resistivity. What is not in general true is that each type of disorder influences the FWHM and the thermal conductivity in the same fashion. There may be certain types of disorder that couple strongly to the optical

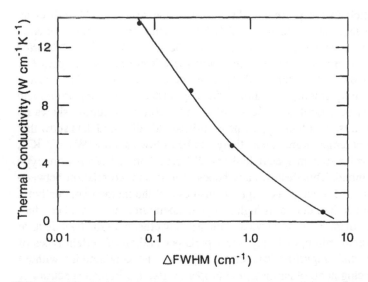

FIG. 29 Thermal conductivity at room temperature of neutron-irradiated single-crystal diamond as a function of the full width at half maximum of the diamond Raman line. Adapted from Ref. 57.

phonons and thus produce significant broadening of the Raman line while influencing the conductivity hardly at all; on the other hand, there may be defects that couple strongly to the heat-carrying phonons and thus provide significant resistivity but do not couple at all to the optical phonons, thereby leaving the FWHM unchanged. One, therefore, must conclude that since there are many types of disorder in CVD films that can affect the thermal and optical properties in different ways, it is too naive to draw a one-to-one correspondence between a given FWHM and a corresponding value of the thermal conductivity that is valid for any CVD diamond.

Although nondiamond carbon can be one of the major defects limiting the thermal conductivity of CVD diamond, the most prevalent defects and amongst the strongest scatterers of phonons in this material are grain boundaries. Despite great improvements in the growth process, all CVD diamond materials grown to date (excluding those grown on diamond itself) have been polycrystalline. The nature of this polycrystallinity and its influence on heat transport is the subject of the next section.

C. Crystallite Size Effects and Anisotropy

In 1985 Nepsha et al. [96] pointed out that if a diamond crystal were reduced to thicknesses well below 1 mm, the thermal conductivity at room temperature and even higher would show a significant size effect, assuming all other scattering processes were the same as in the bulk. Even though this theoretical study dealt with the case of a single crystal in which one dimension is made small, the results would also apply to a polycrystalline film, with the grain size just replacing the smallest sample dimension. Figure 30 shows the expected dependence of κ on grain size according to the Nepsha model.

Experimental evidence that boundary scattering was indeed important for diamond films with grain sizes in the range of a few microns was provided in 1988 by Morelli et al. [95], whose results were described in Section IV.B. They found that the shape of the thermal conductivity curve above 30 K of two films was qualitatively in agreement with that expected due to grain boundary scattering with a grain size of a few

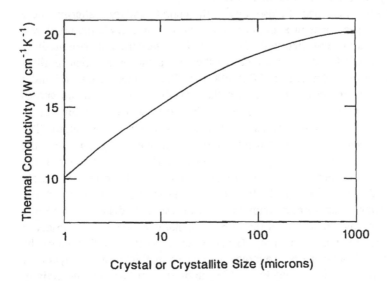

Crystal or Crystallite Size (microns)

FIG. 30 Thermal conductivity at 300 K as a function of crystal size (for single crystal) or crystallite size (for polycrystal) for natural single-crystal diamond. Adapted from Ref. 96.

microns. This feature of the thermal conductivity of diamond films was subsequently verified by other studies [70,98,99] that measured the temperature dependence of κ down to low temperatures.

Graebner et al. [100] suggested that, since grain boundaries are one of the strongest scatterers of phonons and the grain boundary structure of diamond films is columnar, one should expect an anisotropic κ with respect to the film plane. Graebner et al. [101] provided evidence for this by measuring the thermal conductivity of a single film both parallel and perpendicular to the film's surface. The thermal conductivity along the film plane was measured with a standard steady-state technique, whereas that perpendicular to the plane was performed using a laser flash technique that provided a lower limit for κ. The film had the columnar grain structure typical of CVD diamond, with the grain size at the substrate surface less than 1 μm and that at the growth surface on the order of 35 μm. These authors found that conductivity perpendicular to the film was as much as 50% higher than along the film if proper account was taken of the high-resistance layer at the film/substrate interface, and they attributed the result to the predominance of grain boundary scattering in the presence of the columnar structure.

Further evidence of anisotropy in the grain boundary scattering and a resulting gradient in κ through the thickness of a CVD diamond was provided by Graebner et al. [102,103]. They measured the conductivity both parallel and perpendicular to the film plane on a series of films ranging in thickness from 28 to 408 μm. These films were made under identical growth conditions with the growth interrupted at the appropriate time to provide a film of the desired thickness. Graebner et al. claim that the resulting films are identical except for grain structure. The grains were once again shown to be columnar, with the grain size at the growth surface depending on film thickness. The authors found that the conductivity perpendicular to the layers increased from approximately 10 to 20 W cm^{-1} K^{-1} as the film thickness increased from 28 to 408 μm, implying a nonuniform conductivity with respect to distance z from the bottom (low grain size) of the film. A similar dependence was observed for the in-plane conductivity. Graebner et al. pointed out that, since the measured value of κ is an average over the entire thickness, the layer near the top for the film of thickness 408 μm must be of thermal conductivity higher than 20 W cm^{-1} K^{-1}. They analyzed the results in terms of a model in which each additional layer added to the film contributed a conductance that was added to all the other layers and defined a local conductivity κ_1 such that the observed conductivity for a film of thickness Z was expressed as

$$\kappa_{ob}(Z) = (1/Z) \int \kappa_1(z) \, dz \qquad (20)$$

From this they derived the local conductivity versus distance from the bottom of the film; see Fig. 31. Graebner et al. concluded that the value of $\kappa_1 \simeq 21$ W cm^{-1} K^{-1} for the local conductivity at the surface of the 408-μm thick film indicates that near the growth surface the conductivity of this film is at least as high as single-crystal diamond.

V. FINAL REMARKS

The thermal conductivity of diamond is one of the most important physical parameters of this fascinating and potentially technologically important form of carbon. We have seen that for single crystals the experimental information extend from below 1 K to in excess of 1000 K.

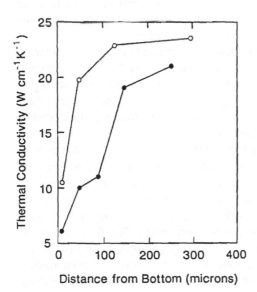

FIG. 31 Local thermal conductivity of CVD diamond films versus distance from the bottom (substrate) side of the film, according to Ref. 103. Open circles, local conductivity perpendicular to the film plane; closed circles, local conductivity parallel to the film plane.

In the case of high-purity diamond, even though a basic understanding of the mechanisms that limit the thermal conductivity is in hand, many outstanding problems with regard to the transport of heat by phonons in this material remain to be answered. A first principles calculation of the strength of phonon–phonon umklapp processes and their dependence on frequency and temperature still eludes us. Whether normal processes are strong enough to explain the change in conductivity upon isotopic purification or whether the observed effects are due to the elimination of other defects as well remains an open question requiring further theoretical and experimental investigation.

In the case of type I diamond, the role of nitrogen in limiting the thermal conductivity still is not entirely clear. Optical studies over the past decade have provided a better understanding of the various forms of nitrogenous defects in these types of diamonds, but experiments aimed at determining the effect on the thermal conductivity of a single type of nitrogen defect to a large extent have not yet been performed. In this regard, it is of interest to study diamonds in which predominantly just one form of nitrogen defect is present as determined optically. Such experiments complement the optical studies and can provide information such as nitrogen aggregate size and concentration.

The case of IIb diamond is an interesting one because of its potential use as a semiconductor device element, yet very little information is available on its thermal properties. The role of electron scattering or point-defect scattering from acceptor states has hardly been explored and can provide important information on the concentration and distribution of carriers and impurities in this material.

Finally, the last 5 years have seen a tremendous improvement in the quality of synthetically produced CVD diamond. This material has evolved from thin, almost black films of a few microns thickness and containing many defects and impurities to large optically transparent specimens upward to 1 mm thick, with grain sizes approaching 100 μm and containing very few defects and impurities. The thermal conductivity of these materials has correspondingly improved greatly as well. The influence of disordered carbon and vacancies on the heat conduction process has been clarified by comparison with neutron- and electron-irradiated single crystals. The important role of grain boundaries in determining the magnitude and anisotropy of conductivity has also been demonstrated. In fact, the growth process has improved to such a large extent over the past 5 years that one can even consider the possibility that the quality of CVD diamond may soon *exceed* that of natural single crystals, which by virtue of their long growth period (on the order of 10^9 years) almost always contain dislocations, stacking faults, nitrogen, and even irradiation-induced defects. Thus, somewhat ironically perhaps, CVD diamond may ultimately prove to be the final testing ground for determining the intrinsic thermal conductivity of the best heat conductor known to man.

ACKNOWLEDGMENTS

I thank Dr. Thomas Perry for illuminating discussions on the optical properties of diamond and Dr. Joseph Heremans for a critical reading of the manuscript. Any remaining errors are the responsibility of the author.

REFERENCES

1. J. C. Angus and C. C. Hayman, *Science 241*, 913 (1988).
2. R. Ramesham, T. Roppel, C. Ellis, and M. F. Rose, *J. Electrochem. Soc. 138*, 1706 (1991).
3. D. G. Blanchard, E. E. Marotta, and L. S. Fletcher, *American Institute of Aeronautics and Astronautics Report No. 92-0708*, AIAA, Washington, DC, 1992.
4. R. Berman, *Thermal Conduction in Solids*, Clarendon, Oxford, 1976.
5. P. G. Klemens, in *Solid State Physics*, Vol. 7, (F. Seitz and D. Turnbull, eds.), Academic Press, New York, 1958, p. 1.
6. P. Carruthers, *Rev. Mod. Phys. 33*, 92 (1961).
7. A. Eucken, *Ann. Physik 34*, 185 (1911).
8. P. Debye, in *Vortrage Uber die Kinetische Theorie der Materie und Elektrizitat*, Teubner, Berlin, 1914.
9. R. Peierls, *Ann. Physik 3*, 1055 (1929).
10. G. Leibfried and E. Schlomann, *Nachr. Akad. Wiss. Gottingen IIa4*, 71 (1954).
11. C. L. Julian, *Phys. Rev. 137*, A128 (1965).
12. P. G. Klemens, in *Thermal Conductivity*, Vol. 1, (R. P. Tye, ed.), Academic Press, London, 1969, p. 1.
13. G. A. Slack, in *Solid State Physics*, Suppl. 34, (F. Seitz and D. Turnbull, ed.), Academic Press, New York, 1979, p. 128.
14. C. Herring, *Phys. Rev. 95*, 954 (1954).
15. C. J. Glassbrenner and G. A. Slack, *Phys. Rev. 134*, A1058 (1964).

16. J. Callaway, *Phys. Rev. 113*, 1046 (1959).
17. R. Berman, *Proc. Roy. Soc. (London) A 208*, 90 (1951).
18. R. Berman, E. L. Foster, and J. M. Ziman, *Proc. Roy. Soc. (London) A 237*, 344 (1956).
19. G. K. White and S. B. Woods, *Can. J. Phys. 33*, 58 (1955).
20. H. M. Rosenberg, *Proc. Phys. Soc. (London) A 67*, 837 (1954).
21. G. A. Slack, *Phys. Rev. 105*, 829 (1957).
22. I. Pomeranchuk, *J. Phys. USSR 6*, 237 (1942).
23. P. G. Klemens, *Proc. Phys. Soc. (London) A 68*, 1113 (1955).
24. T. H. Geballe and G. W. Hull, *Phys. Rev. 110*, 773 (1958).
25. R. Berman and J. C. F. Brock, *Proc. Roy. Soc. (London) A 289*, 66 (1965).
26. R. L. Sproull, M. Moss, and H. Weinstock, *J. Appl. Phys. 30*, 334 (1959).
27. A. Granato, *Phys. Rev. 111*, 740 (1958).
28. A. C. Anderson and M. E. Malinowski, *Phys. Rev. B 5*, 3199 (1972).
29. R. O. Pohl, *Phys. Rev. Lett. 8*, 481 (1962).
30. J. W. Schwartz and C. T. Walker, *Phys. Rev. 155*, 969 (1967).
31. H. B. G. Casimir, *Physica 5*, 495 (1938).
32. J.-P. Issi, *Aust. Jour. Phys. 32*, 585 (1979).
33. R. Berman, *Proc. Phys. Soc. (London) A 65*, 1029 (1952).
34. G. B. Sutherland, D. E. Blackwell, and W. G. Simeral, *Nature 174*, 901 (1954).
35. W. Kaiser and W. L. Bond, *Phys. Rev. 115*, 857 (1959).
36. E. C. Lightowlers and P. J. Dean, *Diamond Research 1964*, Industrial Diamond Information Bureau, London, 1964, p. 21.
37. G. S. Woods, G. C. Purser, A. S. S. Mtimkulu, and A. T. Collins, *J. Phys. Chem. Sol. 51*, 1191 (1990).
38. W. V. Smith, P. P. Sorokin, I. L. Gelles, and G. J. Lasher, *Phys. Rev. 115*, 1546 (1959).
39. R. Berman, P. R. W. Hudson, and M. Martinez, *J. Phys. C 8*, L430 (1975).
40. R. Berman and M. Martinez, *Diamond Research 1976*, Industrial Diamond Information Bureau, London, 1976, p. 7.
41. G. A. Slack, *J. Phys. Chem. Sol. 34*, 321 (1973).
42. E. A. Burgemeister, *Physica 93B*, 165 (1978).
43. J. W. Vandersande, C. B. Vining, and A. Zoltan, *Proceedings of 2nd International Symposium on Diamond Materials*, Electrochemical Society, Pennington, NJ, 1991, Vol. 91-8, p. 443.

44. D. G. Onn, A. Witek, Y. Z. Qui, T. R. Anthony, and W. F. Banholzer, *Phys. Rev. Lett. 68*, 2806 (1992).

45. W. J. deHaas and T. Biermasz, *Physica 5*, 47 (1938).

46. R. Berman, F. E. Simon, and J. M. Ziman, *Proc. Roy. Soc. A 220*, 171 (1953).

47. J. W. Vandersande, *Phys. Rev. B 13*, 4560 (1976).

48. F. C. Frank, B. R. Lawn, A. R. Lang, and E. M. Wilks, *Proc. Roy. Soc. A 301*, 239 (1967).

49. J. W. Vandersande, *J. Physique C6*, supp. 8, 1017 (1978).

50. A. K. McCurdy, H. J. Maris, and C. Elbaum, *Phys. Rev. B 2*, 4077 (1970).

51. R. Berman, E. L. Foster, and H. M. Rosenberg, in *Defects in Crystalline Solids*, The Physical Society, London, 1954, p. 321.

52. E. A. Burgemeister and C. A. J. Ammerlaan, *Phys. Rev. B 21*, 2499 (1980).

53. E. A. Burgemeister, C. A. J. Ammerlaan, and G. Davies, *J. Phys. C 13*, L691 (1980).

54. J. Koike, T. E. Mitchell, and D. M. Parkin, *Appl. Phys. Lett. 59*, 2515 (1991).

55. J. W. Vandersande, *Phys. Rev. B 15*, 2355 (1977).

56. J. Koike, D. M. Parkin, and T. E. Mitchell, *Appl. Phys. Lett. 60*, 1450 (1992).

57. D. T. Morelli, T. A. Perry, and J. W. Farmer, *Phys. Rev. B 47*, 131 (1993).

58. E. R. Vance, *J. Phys. C 4*, 257 (1971).

59. S. D. Smith and J. R. Hardy, *Phil. Mag. 5*, 1311 (1960).

60. G. J. Dienes and D. A. Kleinman, *Phys. Rev. 91*, 238 (1953).

61. H. M. Strong and R. H. Wentorf, *Naturwiss. 59*, 1 (1971).

62. W. Banholzer, T. Anthony, and R. Gilmore, *Proceedings of 2nd International Conference on New Diamond Science and Technology*, Materials Research Society, Pittsburgh, 1991, p. 857.

63. T. R. Anthony, W. F. Banholzer, J. F. Fleischer, L. Wei, P. K. Kuo, R. L. Thomas, and R. W. Pryor, *Phys. Rev. B 42*, 1104 (1990).

64. D. T. Morelli, G. W. Smith, J. Heremans, W. F. Banholzer, and T. R. Anthony, *Proceedings of 2nd International Conference on New Diamond Science and Technology*, Materials Research Society, Pittsburgh, 1991, p. 869.

65. G. A. Slack, R. A. Tanzilli, and R. O. Pohl, *J. Phys. Chem. Sol. 48*, 641 (1987).

66. J. W. Bray and T. R. Anthony, *Zeit. Physik B 84*, 51 (1991).
67. K. C. Hass, M. A. Tamor, T. R. Anthony, and W. F. Banholzer, *Phys. Rev. B 45*, 7171 (1992).
68. V. I. Nepsha, V. R. Grinberg, Yu. A. Klyuev, and A. M. Naletov, *Diamond and Related Materials*, *1*, 891 (1992).
69. R. Berman, *Phys. Rev. B 45*, 5726 (1992).
70. J. E. Graebner and J. A. Herb, *Diamond Films and Tech. 1*, 155 (1992).
71. P. G. Klemens, *Nucl. Inst. Meth. B 1*, 204 (1984).
72. T. H. Geballe, *Science 250*, 1194 (1990).
73. P. G. Klemens, *Phys. Rev. 86*, 1055 (1952).
74. G. A. Slack, *J. Appl. Phys. 35*, 3460 (1964).
75. B. K. Agrawal and G. S. Verma, *Phys. Rev. 126*, 24 (1962).
76. L. A. Turk and P. G. Klemens, *Phys. Rev. B 9*, 4422 (1974).
77. T. Evans and C. Phaal, *Proc. Roy. Soc. A 270*, 538 (1962).
78. T. Evans, in *Diamond Research 1973*, Industrial Diamond Information Bureau, London 1974, p. 2.
79. G. Davies, in *Chemistry and Physics of Carbon*, Vol. 13, (P. L. Walker and P. A. Thrower, ed.), Marcel Dekker, New York, 1977, p. 1.
80. C. D. Clark, R. W. Ditchburn, and M. B. Dyer, *Proc. Roy. Soc. A 234*, 363 (1956).
81. J. W. Vandersande, *J. Phys. C 13*, 759 (1980).
82. T. Evans, in *Diamond Research 1978*, Industrial Diamond Information Bureau, London, 1978, p. 17.
83. J. P. F. Sellschop, D. M. Bibby, C. S. Erasmus, and D. W. Mingay, in *Diamond Research 1974*, Industrial Diamond Information Bureau, London, p. 43.
84. V. I. Nepsha, A. M. Naletov, N. F. Reshetnikov, Yu. A. Klyuev, and G. B. Bokii, *Sov. Phys. Dokl. 30*, 855 (1985).
85. J. W. Vandersande, *Diamond Research 1973*, Industrial Diamond Information Bureau, London, p. 21.
86. F. P. Bundy, H. M. Strong, and R. H. Wentorf, in *Chemistry and Physics of Carbon*, Vol. 10 (P. L. Walker and P. A. Thrower, ed.), Marcel Dekker, New York, 1973, p. 213.
87. K. Kurihara, K. Sasaki, M. Kawaradi, and N. Koshino, *Appl. Phys. Lett. 52*, 437 (1988).
88. A. Ono, T. Baba, H. Funamoto, and A. Nishikawa, *Jap. J. Appl. Phys. 25*, L808 (1986).

89. A. Nishikawa, H. Funamoto, A. Ono, and T. Baba, *Proceedings of 1st International Symposium on Diamond and Diamond-like Films,* Electrochemical Society, Pennington, NJ, 1989, Vol. 89-12, p. 524.

90. A. Sawabe and T. Inuzuka, *Thin Solid Films 137,* 89 (1986).

91. J. A. Herb, C. Bailey, K. V. Ravi, and P. A. Dennig, *Proceedings of 1st International Symposium on Diamond and Diamond-like Films,* Electrochemical Society, Pennington, NJ, 1989, Vol. 89-12, p. 366.

92. R. W. Pryor, R. L. Thomas, P. K. Kuo, and L. D. Favo, *Diamond Optics II,* Vol. SPIE-1146 (A. Feldman and S. Holly, ed.), Society of Photo-Optical and Instrumentation Engineers, Bellingham, WA, 1989, p. 68.

93. S. Albin, W. P. Winfree, and B. S. Crews, *J. Electrochem. Soc. 137,* 1973 (1990).

94. K. Baba, Y. Aikawa, and N. Shohata, *J. Appl. Phys. 69,* 7313 (1991).

95. D. T. Morelli, C. B. Beetz, and T. A. Perry, *J. Appl. Phys. 64,* 3063 (1988).

96. V. I. Nepsha, N. F. Reshetnikov, Yu. A. Klyuev, G. B. Bokii, and Yu. A. Pavlov, *Sov. Phys. Dokl. 30,* 547 (1985).

97. V. I. Nephsa, V. R. Grinberg, Yu. A. Klyuev, N. A. Kolchemanov, and A. M. Naletov, *Proceedings of 2nd International Conference on New Diamond Science and Technology,* Materials Research Society, Pittsburgh, 1991, p. 887.

98. T. R. Anthony, J. L. Fleischer, J. R. Olson, and D. G. Cahill, *J. Appl. Phys. 69,* 8122 (1991).

99. D. T. Morelli, T. M. Hartnett, and C. J. Robinson, *App. Phys. Lett. 59,* 2112 (1991).

100. J. E. Graebner, J. A. Mucha, L. Seibles, G. W. Kammlott, and F. Santiago, *Bull. Am. Phys. Soc. 36,* 646 (1991).

101. J. E. Graebner, S. Jin, G. W. Kammlott, B. Bacon, and L. Seibles, *J. Appl. Phys. 71,* 5353 (1992).

102. J. E. Graebner, S. Jin, G. W. Kammlott, J. A. Herb, and C. F. Gardinier, *Appl. Phys. Lett. 60,* 1576 (1992).

103. J. E. Graebner, S. Jin, G. W. Kammlott, J. A. Herb, and C. F. Gardinier, *Nature 359,* 401 (1992).

3

Chemistry in the Production and Utilization of Needle Coke

Isao Mochida

Kyushu University, Kasuga, Fukuoka, Japan

Ken-ichi Fujimoto

Nippon Steel Chemical Company, Ltd., Kitakyushu, Fukuoka, Japan

Takashi Oyama

Koa Oil Company, Ltd., Kuga-gun, Yamaguchi, Japan

111

I. INTRODUCTION

A billion tons of iron and steel have been produced every year in the world to meet the demand of the highly civilized society of the iron age. Such an enormous production leads to huge amounts of scrap to be regenerated and recycled. Electric arc furnaces are principally operated for this purpose. These furnaces operate under very severe conditions and require graphite electrodes of high performance to reduce the cost of steel making by stable operation over a long term.

The use of gas-powered engines worldwide produces a large demand for gasoline, but recent crude runs tend to give heavier and heavier fractions (higher boiling range) that must be upgraded thermally or catalytically to fuel oil of a certain boiling range and purity. Thermal cracking processes convert the heavy fractions into distillate and coke; the distillate is enriched in hydrogen, whereas the coke produced is low in hydrogen. Delayed coking is one of the most popular thermal cracking processes, being operated commercially all over the world. The coke produced from the delayed coker is very frequently regarded as an unwanted by-product, which is fired as a very cheap solid fuel. However, careful selection of feedstocks and cracking conditions allows production of a valuable carbon source, needle coke, for making graphite electrodes for electric arc furnaces and other wide applications. The name of *needle coke* comes from its appearance. The coke appears to consist of bunches of carbon needles that are aligned in one direction as shown in Fig. 1.

The contrast in features is clear by comparing needle coke structure to that of regular mosaic coke. Such a laminar structure ensures that the coke will satisfy properties as the filler of a high-quality electrode.

Needle coke (Fig. 2) can be produced from coal tar as well as from petroleum. Needle coke from coal tar has enlarged the coal tar business, and hence coal utilization. Thus, the delayed coking, by producing distillate and coke of high value, allows the efficient utilization of the bottom of barrel of our precious fossil resources, although some pretreatment before coking is often necessary.

Figure 3 shows the current consumption of graphite electrodes in the world [1]. More than one million tons of electrode were consumed every year in the 1970s. Drastic decreases were experienced in the 1980s because of the worldwide depression after the oil crisis. Although some recovery was observed in 1984, the consumption decreased again in 1986 because the advanced technology for better electrodes (mainly due to better coke) and furnace operation reduced significantly the graphite consumption per unit steel production. Figure 4 shows world steel

FIG. 1 Needle and regular cokes.

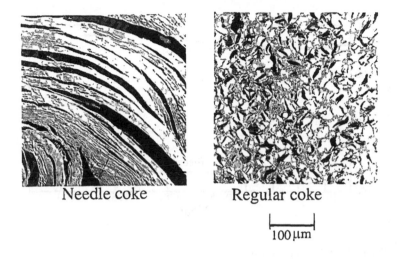

Needle coke   Regular coke

100 μm

FIG. 2   Texture of needle and regular cokes.

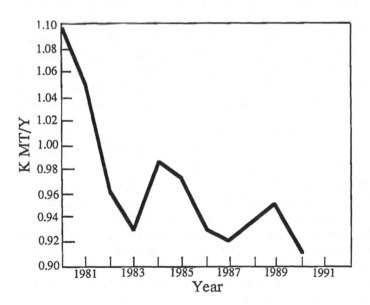

FIG. 3   World graphite electrode consumption.

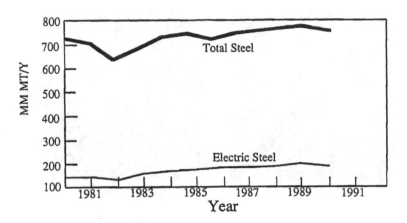

FIG. 4 World steel production.

production [1]. Through the 1980s, the total steel consumption in the world may increase steadily and electric steel may increase its share in the developed countries because of rapid growth of the use of gas-powered engines. Hence, the demand for needle coke can keep or slightly increase the present level of its production, although a rather sharp reduction was forecast in 1990 by Sooter [1]. In any case, needle coke of better and more stable quality as well as lower cost is continuously searched for.

In the present review, the chemistry of the production and utilization of needle coke is surveyed. The review covers the properties and structure of needle coke, commercial production procedures for needle coke, and carbonization chemistry specific to the needle coke production, extrusion, baking, and graphitization of needle coke into the electrode. Most of the emphasis is put on the chemistry of coke and carbonization. The chemistry discussion is aimed at understanding chemical schemes and mechanisms for the production of highly graphitizable needle coke in the very specific structure that produces a desirable very low coefficient of thermal expansion (CTE). Such a mechanistic understanding may suggest future routes for better coke at lower cost from any feedstocks available.

II. GRAPHITE ELECTRODES

A. Needle Cokes for Electrodes

A graphite electrode in an electric arc furnace for steel production is usually a large cylinder, 0.1–0.6 m in diameter and 2 m long, as shown in Figure 5. It conducts electricity, which heats the metal through an arc in the furnace.

The electrodes are made from two major raw materials; needle coke, which fills the electrode, and binder pitch, which helps the cokes to be extruded into the "green" or not yet graphitized electrode and bound together during the baking and graphitization processes. The raw materials used in the production of the electrode essentially determine its ultimate properties. Electrodes for ultra-high power (UHP) arc furnaces have to withstand very severe electrical and mechanical conditions as one might imagine from Fig. 6. This requires a high electrical conductivity and high thermal shock and oxidation resistance.

Ideally, the needle coke of the electrode should have a low sulfur and nitrogen content, because these atoms cause "puffing" or uncontrolled irreversible expansion during graphitization, which leads to lower

FIG. 5 Graphite electrode and nipple.

FIG. 6   Electric arc furnace using electrodes.

strength of the electrode. Low CTE is a most important requirement because high thermal shock resistance of the electrode is directly related to the low CTE of the filler cokes. Details of electrode manufacture, testing, and applications are available [2–8].

B. Procedures of Electrode Manufacture

Figure 7 shows the flow sheet for the manufacture of a graphite electrode, which consists of several steps.

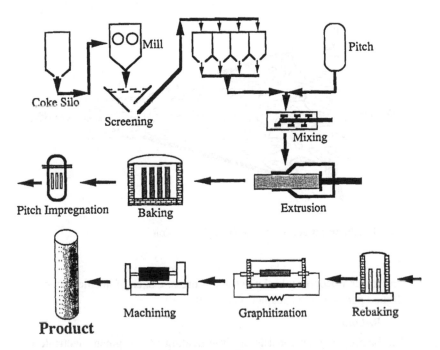

FIG. 7 Flow sheet for manufacture of graphite electrode.

1. Preparation of Raw Materials

After crushing, the calcined needle cokes are screened and classified according to their grain sizes to prepare the optimum size composition for the electrode. The grain size distribution is very important for the strength and CTE of the electrode. The fractions of each grain size are stored in individual vertical storage hoppers. The weighed fractions of the coke and binder pitch are mixed at a temperature between 160 and 180°C until a homogeneous paste is obtained. After cooling to 100°C, the blend is introduced into an extrusion press, which produces the "green stock."

2. Forming (Extrusion)

The hot plastic paste of filler coke and binder pitch is forced through dies at about 120°C to form an electrode of suitable length and cross section. Figure 8 illustrates the macroscopic alignment of coke fillers in the

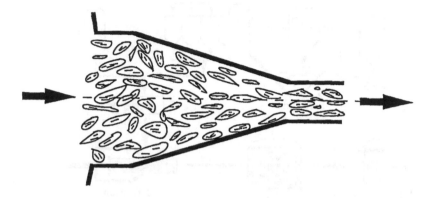

FIG. 8 Alignment of needle coke through extrusion.

binder pitch occurring in the extrusion press. Cokes of needle shape are aligned parallel to the adapter wall [4,9].

3. Baking

The baking process consists of slow heating of the green electrode to 800–1000°C and holding for a few days at that temperature. This step converts the thermoplastic pitch into a solid coke, while the extruded electrode keeps its shape. The content of binder pitch and heating rate are very important for this step. Another function of slow baking is to minimize the shrinkage associated with pyrolysis of the pitch. The conversion of the pitch to coke on the needle coke surface is accompanied by marked physical and chemical changes, which, if conducted too rapidly, lead to serious quality faults in the final product such as expansion, distortion, and pits. For this reason, baking is generally regarded as the most critical operation in the production of the graphite electrode. The resulting carbon electrode is called a *baked electrode*.

4. Impregnation

The idea of impregnation is to increase the density of electrode by filling the open pores of the baked electrode with impregnation pitch. These pores are produced inevitably by the devolatilization from the binder pitch during the baking step. The impregnation pitch is usually quinoline-insoluble (QI) free pitch of high coking value. The first step in the impregnation process is to evaluate the electrode. Melted pitch is introduced and covers the electrode thus filling its pores. The pressure is increased to about 10 atm to shorten the time for pitch penetration. For UHP furnaces, which require high-strength electrodes, this impregnation process is necessary. Impregnated electrodes are rebaked at 800–1000°C with slow heating for several hours. The outflow of impregnated pitch from the electrode during rebaking is a major problem. The impregnation and rebaking are repeated several times when very high density electrodes are required.

5. Graphitization

The baked or rebaked electrode is then graphitized by heating it to 3000°C. The purpose of this step is to convert the carbons from the filler coke and binder pitch into graphite from which the highest strength, chemical restivity, and electric conductivity are obtained. The furnace for graphitization invented by Acheson is still in use today with only minor modifications. It is an electrically fired furnace capable of heating electrodes of several tons to a temperature approaching 3000°C. Recently, lengthwise graphitization has been introduced by some electrode manufacturers. In this process, electric current is passed directly through the baked electrode, generating within the carbon the heat required to raise it to graphitization temperatures. Lengthwise graphitization allows rapid heating, uniform temperature distribution, and shorter graphitization time, thus also saving energy. During graphitization, the electrode must have good dimensional stability. If puffing caused by evolution of sulfur and nitrogen compounds during graphitization occurs, the electrode may suffer formation of cracks and fissures or a decrease of its bulk density. The faster the heating rate, the more puffing problems can arise.

C. Commercially Available Electrodes

The diameter of commercial electrodes has increased progressively because of the greater current-carrying capabilities demanded by large, high-power arc furnaces. Common diameters of today's electrodes are 610 mm. The maximum length of commercial electrodes is currently 2800 mm. Besides electrodes, nipples used as connecting pins for the electrodes are also produced. These nipples are produced by slightly different procedures than that of the electrodes because they need greater electrical conductivity and high mechanical strength. Finer coke grains and more impregnation and rebaking cycles are required.

TABLE 1   Properties of Graphite Electrode

	Regular	UHF
Bulk density (g/cm^3)	1.55–1.70	1.60–1.75
Real density (g/cm^3)	2.20–2.21	2.20–2.22
Electrical resistance ($\mu\Omega$m)	6.0–10.0	4.0–7.0
Young's Modulus (kg/mm^2)	700–1000	800–1400
Bending strength (kg/mm^2)	80–160	110–170
CTE (10^{-6}/°C)	0.7–2.7	0.3–1.0
Ash (%)	0.2	0.2

Table 1 summarizes the specifications of the commercial graphite electrode [6]. For UHP furnaces, electrodes with high electrical conductivity and high thermal shock resistance are strongly desired.

D. Electrode Consumption and Thermal Shock Resistance

The principal causes of electrode consumption in the furnace are (a) evaporation of graphite at the tip of the electrode, (b) oxidation of carbon on the side surface, (c) breakage of the joints due to excess mechanical stress, and (d) breakage due to thermal stress.

The major part of the tip consumption is due to vaporization of the graphite in the hot spot, where the temperature reaches between 3600 and 4000°C. Reduction of tip consumption is achieved by using lower temperatures and smaller temperature gradients at the top, which is done by reducing the electric current level. Electrodes used under these milder conditions do not have to go through the impregnation process during manufacture, since regular grades can adequately meet the properties required under these less severe conditions.

Oxidation causes side wall consumption. Although oxidation of the electrode surface is almost negligible below 500°C, it is very rapid above 1000°C; the outer surface of the electrode is principally oxidized because diffusion of the oxidant is relatively slow. Side wall consumption is reduced by decreasing the total surface area of the electrode exposed to the furnace atmosphere (air) and by shortening the duration of the melting and refining stages (i.e., the tap-to-tap time, which controls the contact time in air).

Breakages (c) and (d) are classified as intermittent consumption. Intermittent consumption is due to electrode breakage, portions of which remain in the furnace and cannot be reused. Thermal and mechanical stresses on the electrodes are mainly responsible for such breakages. The breakage almost always occurs at the point between jointed parts of the electrode.

A number of studies revealed the influence of electrode quality of specific electrode consumption during steel production [2]. The following properties were suggested as being important: electrical resistance, real density, and mechanical (bending) strength. The best performance of the electrode is normally associated with its low electrical resistance and high mechanical strength. Thermal shock resistance (R), expressed by Eq. (1) is believed to be one of the most important factors for defining the quality of the electrode during its working state because the large-diameter electrode suffers a very high thermal shock at the peak load in the furnace.

$$\text{Thermal shock resistance}, R = S \times k/E \times \alpha \qquad (1)$$

where S is bend strength, k is thermal conductivity, E is Young's modulus, and α is the coefficient of thermal expansion. Higher strength, thermal conductivity, lower modulus, and lower CTE are critically necessary for high thermal shock resistance. Compared with other materials, graphite is outstanding in resistance to thermal stress as well as high temperatures above 1000°C because of its very balanced properties. Hence, filler cokes should have low CTE, high density, and low puffing tendency because puffing decreases the bulk density and hence strength of the electrode. It is also important for the coke to have adequate affinity for the binder pitch.

Oxidation, which is the only disadvantage of graphite, decreases the strength of the electrode in addition to oxidative consumption. However, oxidation at very high temperature is limited to the outer surface. Thus, for very dense electrodes with few pores, little reduction of strength is caused by oxidation.

The sudden temperature change in the furnace certainly increases thermal shock. Hence, the operation control of the furnace is another key factor for the electrode consumption.

III. COMMERCIAL PRODUCTION OF NEEDLE COKE

A. Delayed Coking

Needle coke is produced through two steps of delayed coking from petroleum residual oils and coal tar and calcination of the green coke in the kiln. Figure 9 illustrates the delayed coking process. The selected and prepared feed, sometimes blended, is heated in a tubular preheater in a furnace very rapidly to the coking temperature and then transferred to the bottom of a coking drum where the heated feed starts to carbonize without additional heat supply in the thermally insulated drum. The separation of the heating and coking zones is the purpose of the delayed coking. It is very important for the operation that no coking takes place at all in the heating tube, which is of much narrower diameter than the coker drum, in order to avoid any plugging problems. The constituents of feed, its flow rate, and the temperature, which should be higher than that needed for coking because no heat is supplied to the large diameter drum, must all be carefully designed and controlled. Steam or other gases are often

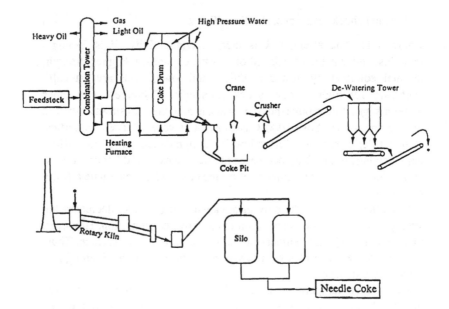

FIG. 9   Delayed coker process flow diagram.

supplied to maintain a rapid flow rate. The feed in the drum loses heat because of losses from the drum wall and endothermic cracking (dealkylation and paraffin cracking) reactions. Volatiles produced by the cracking are removed from the top of the drum. The cracked product is distilled into naphtha, gas–oil, and heavy distillate for recycle to the coker. The heavy hydrocarbons staying in the drum continue their cracking, this time exothermic condensation of aromatic constituents being more emphasized, and are gradually converted into coke through the intermediate mesophase state. Heated feed is continuously supplied to the drum to form stages of coking layers that undergo their respective coking reactions as illustrated in Fig. 10. The drum gradually reaches constant temperature through heat input from the feed and evolution from the progress of endothermic reactions. It should be noted that the feed is always mixed with the recycled distillate, produced from the lower part of the drum. Highly aromatic constituents of relatively small size produced from the feed during coking play some role in moderating

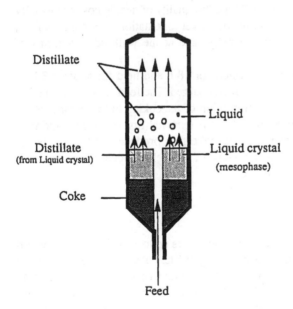

FIG. 10 Coke formation in coke drum.

the carbonization as described in the later sections, especially when the feed is fairly low in aromaticity.

The cokes produced in the drum can be very different in their structure and properties depending on the location of their production. The coke structure will depend primarily on the factors already described, but other factors are also influential. The coking reactions are very much influenced by the flow of evolved and feed gases, which act to strip out light products and provide an inward flowing force. The gases pass the layers of coking substances principally through the center of the drum, especially when the layers reach the last stage of coking. The traces of gas flow are very difficult. Nevertheless they influence the coking reactions strongly and delicately depending on the feed and the coking conditions. The feed to the drum forms layers of coke in progressive stages. The upper layers increase the pressure on the coking lower layers. Thus, the coking temperature and pressure, as well as feed constituents, in a very exact sense, are variable depending on the height and levels in the drum. Heat release from the surface of the drum leads to the temperature distribution along the diameter in the drum because the high viscosity of the layer at the last stage of the coking prevents mixing. Thus, the quality of needle cokes is usually poor at the bottom in terms of lower orientation but high density, at the top in terms of low density, and at the wall sides in terms of low orientation.

The coke drum is usually filled for 18 h and held for another 8 h to complete the coking of whole feed charged. The feed is then switched to another drum. Before the completion of the coking in the second drum, the coke produced in the first drum is removed first by drilling and then by cutting with a water jet. The coke thus produced and recovered is called *green*. An excellent review of the delayed-coking process was given by R. DeBiase et al. [10].

B. Calcination

The green coke is calcined at temperatures up to 1400°C in a rotary kiln or hearth after drying, as shown in Fig. 9. The volatile content of the coke is reduced to a fixed percent. The heating rate, final temperature, heating atmosphere, and amount of volatile matter during the calcination influence the grain size and detailed structure of the resultant coke, hence defining its quality as discussed later [11].

The calcined coke is sold to electrode and carbon manufacturers after sieving. The grain size distribution is an important specification.

C. Coking Feed

The feeds for premium needle cokes today are vacuum residues of low-sulfur crude, tar prepared from the pyrolysis of vacuum gas–oils of low-sulfur crudes, decant oils from the fluid catalytic cracking (FCC) process, steam-cracked tar from the naphtha cracking process for ethylene manufacture, and coal tar. Blending of petroleum tars and/or pretreatments are essential for the production of premium coke. The contents of sulfur and nitrogen in coking feeds should be as low as possible. These elements interfere with the coking reactions, and some stay in the product coke as deleterious impurities.

Figure 11 illustrates several routes for needle coke production from petroleum residues. There are four major procedures depending on the starting feedstocks.

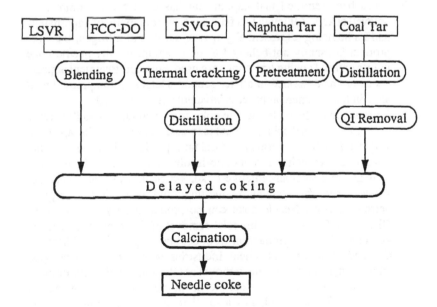

FIG. 11   Production process flow sheet of needle coke.

1. Vacuum residue of low-sulfur crude (LSVR) and FCC-decant oil (FCC-DO) are blended at a suitable ratio for cocarbonization. Careful selection of the crude, adequate FCC operation to achieve a stable quality of decant oil, and extensive removal of salts and cracking catalyst particles are required. (The heaviest fraction of cracked oil is usually recycled.) Contaminants of vacuum residue (VR) tend to accumulate in the coke. Decant oil from hydrodesulfurized gas–oil can be used and may be prepared depending on its sulfur and nitrogen contents. Vacuum residues or decant oils alone have been fed to delayed cokers. The former produces regular or anode coke, and the latter gives needle cokes with high CTE, as described later.

2. Vacuum gas–oil of low-sulfur crude (LSVGO) is pyrolyzed and the vacuum residue of its pyrolyzed product is carbonized with the VR of the same crude. The principal source of the coke is believed to be the former residue; hence, the contaminants in the resulting coke product are usually less than in the coke produced through the procedure described previously. This is because VGO contains fewer contaminants than the VR of the same crude. Catalytic and pyrolytic cracking may give different structures and properties to the feed for the coker even when the same VGO is used.

3. It has been reported that naphtha tar has been fed to commercial delayed cokers. Although this feed consists principally of naphthalene and its homologues, essentially free of sulfur and nitrogen, the coke product is usually not believed to be premium quality needle coke unless unusual coking operations are used. Some pretreatment appears necessary as described in a later section. Details of the pretreatment used in the commercial process have not been disclosed.

4. A production scheme for needle coke from coal tar is also illustrated in Fig. 11. Delayed coking of coal tar was developed first in Japan and still is unique in the world [12]. Coal tar is produced in a conventional coke oven; contains a quinoline-insoluble fraction, primary; and provides regular nonneedlelike coke, which is principally used for anodes used for aluminum smelting; and is similar to regular petroleum coke. Needle coke can be produced only from primary QI-free tar. Thus, the commercial removal of QI from coal tar is the key to needle coke production from coal tar. Nippon Steel Chemical Co. and Mitsubishi Chemical Industries independently developed antisolvent processes for the removal of QI, using heating oil. Primary QI grains coagulate to precipitate in the poor solvent with the aid of binding forces provided by a small amount of a quinoline-soluble (QS) fraction.

More detailed chemistry on the preparation of feeds including their selection, blending, and pretreatment is described in Sections VII and VIII.

D. Needle Coke Producers in the World

The producers of needle cokes in the world are listed in Table 2, with the feed used and the amount of production. The general properties of commercial needle cokes from petroleum residue and coal tar pitch are summarized in Table 3. Their grading is not simple. Electrode manufacturers may select them by their price, CTE, density (structure), grain size and shape distribution, and puffing extent. Some manufacturers care very much the amount of binder pitch required and viscosity of the slurry at extrusion.

E. Coking Problems

Although a million tons of needle coke are now produced commercially every year, there are several problems to be solved for better quality and lower cost.

1. Stable quality of cokes in terms of drum location, slight changes of feed properties, and day-by-day operation (key controls).
2. Better quality in terms of lower CTE, higher density, grain size distribution, lower impurities, and puffing.

TABLE 2   Producers of Needle Coke

Company	Capacity (1000 ton/y)	Feed stock
Continental Oil	510	Petroleum residue
Airco Carbon	110	Petroleum residue
Koa Oil	100	Petroleum residue
Shell	100	Petroleum residue
Union Oil	100	Petroleum residue
Nippon Steel Chemical	70	Coal tar pitch
Mitsubishi Chemical	60	Coal tar pitch
Petro Coke	60	Petroleum residue
Marathon	60	Petroleum residue
Others	40	
Total	1210	

TABLE 3   Properties of Needle Coke

	Petroleum based	Coal tar based
Total moisture (%)	0.1	0.1
Ash (%)	<0.1	<0.1
Real density (g/cm^3)	2.13	2.13
Size, (%) <5 mm	>40	>40
Sulfur (%)	0.4	0.25
Nitrogen (%)	0.4	0.5

3. Control of surface chemistry in the interaction with binder pitch.
4. Variation of feed quality—its detection control and economical pretreatment.
5. Easier removal of dense coke from the drum.

The detailed mechanistic understanding of the whole coking process will be reviewed in the following sections to explore ways to solve these problems.

IV. STRUCTURE AND PROPERTIES OF NEEDLE COKE

A. Microstructures of Needle Coke

Needle coke is one form of a typical graphitizable carbon; hence, it has a microstructure as illustrated by Franklin [13] and in Fig. 12a. Such a

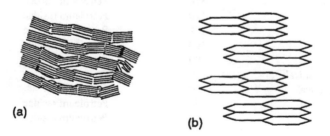

(a)

(b)

FIG. 12   Model of (a) graphitizable carbon and (b) graphite.

microstructure grows easily into the graphitic layers (Fig. 12b) by heat-treating above 2000°C.

The graphitizability of needle coke is an extremely important property, but at present, there is no way to predict the crystallinity of the graphite produced from today's commercial graphitizable needle cokes. There is no distinct relationship between needle coke structure and graphite properties.

The microstructure of green coke is very different from those of calcined and graphitized cokes, green coke being rather similar to the molecular crystals of large condensed aromatic hydrocarbons. The aromatic planes are believed to have layer stacking. The progressive changes in the microstructure of needle cokes during heat treatment at a series of temperatures are illustrated (according to Marsh and Griffiths) in Fig. 13 [14].

The graphitelike microcrystalline units in Fig. 13 are aligned in an approximate orientation in the needle coke and respond to polarized light, as illustrated in Fig. 14, to firm isochromatic (or same reflectance) areas when viewed in the reflection mode of a polarized light

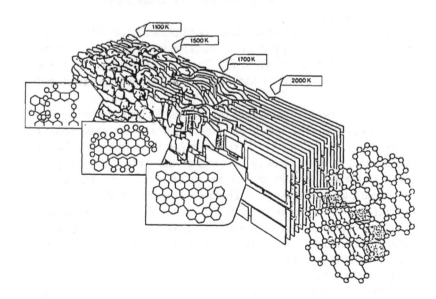

FIG. 13 Model of structures in carbon of HTT 1100–2000 K.

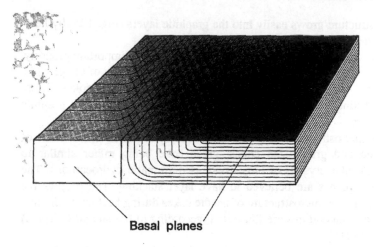

Basal planes

FIG. 14 Color from polished section of carbon.

microscope. The distribution of isochromatic areas forms the optical textures in the coke.

Premium needle coke characteristically exhibits a beautiful flow-type exture where the narrow isochromatic belts run in the same direction for several centimeters, as shown in Fig. 2a, whereas regular coke has a mosaic texture where rather spherical isochromatic areas are distributed as shown in Fig. 2b.

Figure 15 shows micrographs of two perpendicular cross sections of needle cokes [15]. The anisotropic flow textures in the cross section parallel to the axis of the needle coke run parallel to the axis of CTE measurement as often reported. In contrast, the cross section perpendicular to the axis exhibits several groups of concentric symmetry of isochromatic units around a pore. There are a number of disclinations in this surface. The areas away from the pores show a rather random orientation of isochromatic units, suggesting that the pore is the center of symmetry. On the other hand, the two cross sections in the mosaic coke look very similar regardless of the axes.

A grain of needle coke has been reported as showing excellent orientation along the vertical axis. The microscopic examination of a whole grain of coke suggests that the pores are the centers of concentric

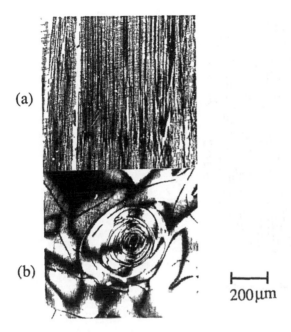

(a)

(b)

200 μm

FIG. 15 Textures of two perpendicular sections of needle coke lump.

symmetry, around which planar disks are oriented in a columnar arrangement. Hence, such a structure of the coke is compared to a bunch of cigars, each of which has its own symmetry center as illustrated in Fig. 16.

It has been widely recognized in the sectioned surfaces of a calcined needle coke that cracks run parallel to the axis of the flow texture as shown in the photomicrographs in Fig. 17.

The structural model shown in Fig. 16 suggests that the cracks will run at the border of the "cigars" since shrinkage takes place during the calcination stage, thus decreasing the diameter of each cigar. Thus, pores, which are believed to be tracks of bubbles produced in the viscous liquid just before solidification during carbonization, are suggested to be intimately related to the formation of the uniaxial orientation of carbon lamellae in the needle coke structure.

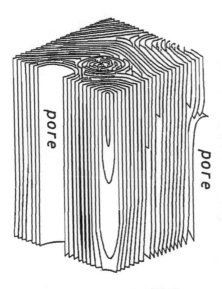

FIG. 16 Structure model of needle coke.

When the entire area of a piece of coke is observed at low magnification under a microscope, it is found to consist of anisotropic (isochromatic) units, each of which has its own size and shape as shown by the montage micrograph of Fig. 18. Thus, the extent of uniaxial orientation of carbon planes in a lump of coke can be evaluated by microscopic point counting of each anisotropic unit.

Figure 19 illustrates how to measure the anisotropic units and to describe the extent of orientation [16].

Table 4 summarizes the extent of orientation of some coke grains prepared at different temperatures. The extent of orientation is also evaluated by quantifying x-ray diffraction intensities by the Bacon method to give the so-called Bacon Anisotropy Factor (BAF), since x-ray diffraction is intensified by increasing orientation [17].

B. Structural Changes in Needle Coke By Heat Treatment

Heat treatment of needle coke between 600 and 3000°C (calcination and graphitization) slightly influences its optical texture under the

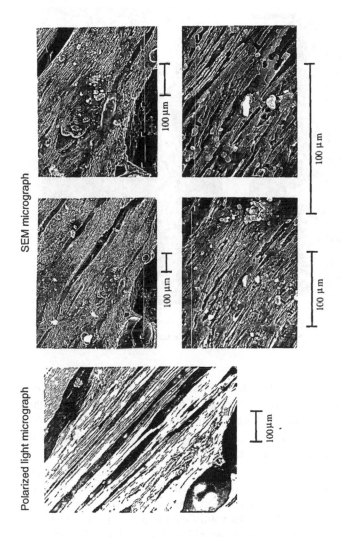

SEM micrograph

Polarized light micrograph

FIG. 17   Texture of needle coke.

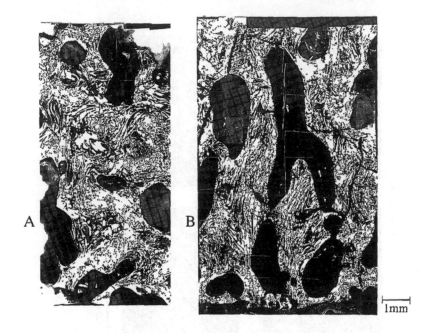

A   B

1mm

FIG. 18 Montage photomicrographs of needle coke.

microscope while graphitization takes place. The graphitization certainly increases the extent of orientation observed by the Bacon method
because the increased crystallinity intensifies the x-ray diffraction. The
optical microscope does not have enough resolution to distinguish the
crystal growth within an anisotropic unit.

The pores and cracks in needle coke are believed to influence properties such as density, strength (hence grain size), reactivity, and CTE.
Voids in the needle cokes are pores and tracks (fissures) that are
produced by gas bubbles at the solidification stage of the coking and
shrinkage due to solidification, graphitization, and cooling (thermal
shrinkage). The gas evolution at the calcination stage is believed to
produce both cracks and pores, whose size and location are widely
distributed. Distribution in their size and location in the coke, as illustrated by Fig. 17, is, in principle, quantifiable under microscopes of
variable resolution, but in practice this is difficult. The large pores do not
change much during heat treatment, however, meso- and micropores and

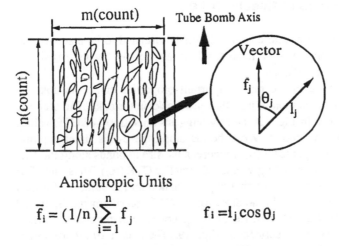

$$\overline{f}_i = (1/n)\sum_{i=1}^{n} f_j \qquad\qquad f_i = l_j \cos\theta_j$$

FIG. 19 Evaluation of anisotropy by montage method where l_j, Length of each anisotropic unit vector (j); n, number of unit vectors in i column; f_i, average f_j in i column parallel to the axis; f_j, the axial component of an anisotropic unit vector in column i; θ_j, angle of an anisotropic vector (j) to the CTE axis.

cracks appear or disappear depending on the considerable shrinkage of the coke due to the progress of graphitization.

It should be noted that crack development and direction of shrinkage are strongly influenced by the optical texture and the orientation of the coke because large shrinkage occurs in the direction perpendicular to the carbon lamellae of the cigar bunch structures in the needle coke [17].

TABLE 4 Relationship Between CTE and BAF of Coke

HTT (°C) CTE ($\times 10^{-6}$/°C)	600	1300	2000	2500
$\alpha_{//}$	10.9	0.4	0.3	0.5
$\alpha_{\perp}$	24.6	2.6	2.4	2.1
BAF	2.7	3.0	5.6	5.1

$\alpha_{//}$, parallel to the preferred orientation.
$\alpha_{\perp}$, perpendicular to the preferred orientation.

C. Properties of Needle Coke

The properties of needle coke are intimately related to its macroscopic as well as microscopic structures, which are defined by its origin and heat-treatment conditions.

The CTE, one of the most important properties of needle coke, is related to the orientation of lamellae and cracks in the coke. The graphite structure shown in Fig. 12b clearly suggests anisotropy of CTE values along the a- and c-axes. Weak bonds along the c-axis (intergraphitic planes) give a larger CTE ($\alpha_\perp$), whereas the strong bonds along the axis (within the plane) should give a much smaller CTE ($\alpha_{//}$). Thus, the CTE of lump cokes depends to a first approximation on the number of a-axis carbon planes oriented in the direction of CTE measurement. Figure 20 correlates the CTE of a series of lump cokes with measures of their anisotropic texture described in Fig. 19. There is a fine correlation, suggesting the importance of flow texture for the value of CTE.

Table 4 summarizes the correlation of CTE with BAF of lump cokes calcined at different temperatures [18]. There is a fair correlation. The

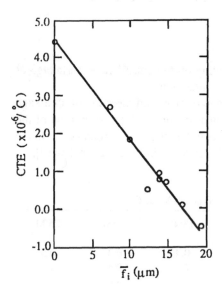

FIG. 20 Relationship between $\bar{f}_i$ and CTE.

CTE values of the lump coke in two perpendicular directions show strong anisotropy, reflecting the anisotropic structure of needle coke.

The CTE parallel to the orientation axis is comparable to that of single crystal graphite parallel to the a-axis [19]. However, it should be obvious that the values corresponding to the c-axis are much smaller than that of single-crystal graphite. There are two reasons, one is a contribution of a-axis planes and the other is microcracks between the planes oriented in the same direction.

The porosity of the coke has been widely reported to be correlated to its mechanical strength as described by Eq. (2), proposed by Knudsen [20]

$$S = S_{max}e^{-bP} \tag{2}$$

where S and S_{max} are the strengths of material having porosity P and zero porosity, respectively, and b is an empirical constant.

More precise correlations will require a more precise quantitative description of voids and a consideration of their locations.

Heat treatment up to graphitization changes the structure and hence the properties of the coke, such as electrical conductivity, CTE, and mechanical strength. Figure 21 illustrates the change of CTE with heat-treatment temperature [21]. Such changes are qualitatively ascribed to the growth of graphite crystals defined by the crystal size and lattice parameters of the graphite and densification due to growth.

V. CHEMISTRY OF EXTRUSION, BAKING, AND GRAPHITIZATION OF NEEDLE COKE

A. Behavior of Needle Coke in Electrode Manufacture

Needle coke sent to the electrode manufacturer is ground and sieved into a particular distribution of grain size, mixed with the minimum amount of binder pitch for extrusion into green electrodes, baked, impregnated, and rebaked. The final three steps are repeated several times, and finally the product is graphitized into the final shape of the graphite artifact as described in Section II. Needle coke is required to meet particular specifications at each step for better performance.

The calcined needle coke should have particular particle size distributions and shape, porosity for sufficient strength, suitable wettability to the binder and impregnation pitches, and low puffing tendency in

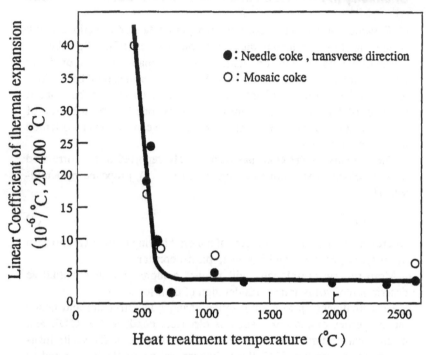

FIG. 21 Changes of CTE with HTT.

addition to a very low CTE. In this section some chemical aspects of extrusion and graphitization of needle cokes are reviewed to clarify what is required to produce high-quality cokes.

B.   Extrusion of Needle Coke-Binder Pitch Slurry

Extrusion of the slurry is requested to produce a green electrode of high density at relatively low temperature with the smallest amount of binder pitch. First of all, a particular distribution of grain size is strictly requested by the respective electrode manufacturers. Figure 22 schematically illustrates the types of volumes of natural cokes, including closed pore (V_c), open pore in the coke (V_p), and voids among the coke grains (V_v), which are the factors for determining the tapped density of the green electrode, as shown by Eq. (3)

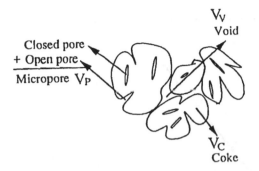

V_V
Void

Closed pore
+ Open pore
Micropore V_P

V_C
Coke

FIG. 22 Schematic coke and pore volume in cylinder where V_C, volume of coke (1/Real Density); V_P, volume of micropores per unit weight of coke (mercury porosity); V_V, volume of voids (including pore exceeding 7500 Å in radius) per unit weight of coke.

$$\text{Tapped density,} \, D = \frac{1}{V_C + V_P + V_V} \qquad (3)$$

where the tapped density of needle cokes is obtained by tap-packing cokes in a cylinder.

Figure 23 indicates the major contribution of tapped density to the bulk density of the green electrode. Hence, tapped density can be a measure of needle coke quality.

Figure 24 illustrates the values of V_c, V_p, and $1/D$. The difference between $1/D$ and $V_c + V_p$ reflects V_v, which is revealed as a major factor in the tapped density. The grain size and shape may govern the packing of cokes in the cylinder [22]. Recently, shape factors of the grains have become important considerations. Needle shapes are believed to hinder smooth packing because of their stable framework. The binder pitch may lubricate the coke surface, partly solving the problem, although details have not been studied.

Another extrusion variable is the viscosity of coke-pitch slurry under extrusion conditions. One needs to obtain an adequate extrusion viscosity with the least amount of binder pitch at the lowest possible temperature. Figure 25 illustrates the torque required for stirring coke-pitch slurries at various temperatures. Although higher temperatures naturally reduce the torque (lower viscosity) regardless of the coke, the coke from coal tar

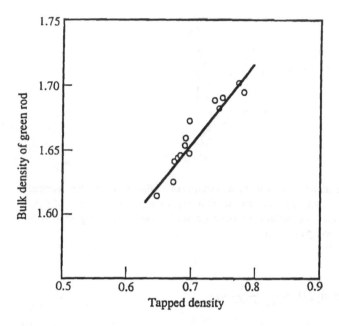

FIG. 23 Relationship between tapped density and bulk density of green rod.

pitch needs a higher torque than that from petroleum residua at the same temperature with the same amount of binder pitch. Stronger adsorption of the binder on the former coke is postulated to account for this. More binder pitch or a higher temperature is required with the former coke, but this is undesirable. It has been found that calcination at higher temperatures or under reducing conditions lowered the required torque, suggesting that stronger interaction of the binder pitch with the coke surface increases the viscosity of the slurry. The importance of the surface structure and chemistry of the coke cannot be underestimated.

Pores that trap the binder pitch may also increase the viscosity of the slurry; thus, fewer macropores are favorable. However, because the pores in the electrode as well as the cokes should be filled to obtain high density in the electrode, it is not a good idea to prohibit the diffusion of pitch into pores at the impregnation stage; better diffusion into pores is strongly desired and greater wettability may be favorable at this stage.

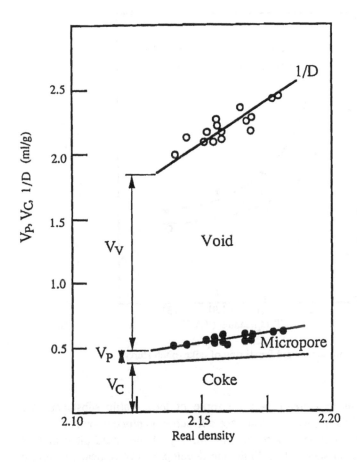

FIG. 24 Composition of green rod.

C. Carbonization in the Pores and/or on the Surface of Coke in the Baking Process

The binder and impregnation pitches are carbonized on the surface or in the pores of the coke in the baking process. The flow, impregnation, and evaporation of the pitch during carbonization are influenced by chemical interaction with the surface of the coke; thus, the yield, porosity, and anisotropic development of the pitch-derived coke is modified. The coke

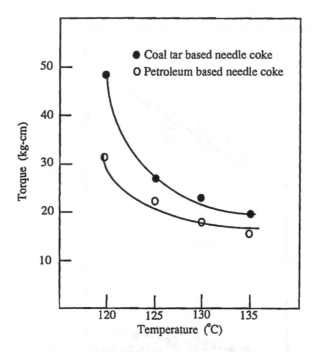

FIG. 25 Torque required for stirring of coke–pitch slurry.

yield tends to increase because some of the volatile substances are chemically adsorbed on the surface and subsequently carbonized.

Such adsorption may reduce the fluidity of the fused pitch and its liquid crystal state, which limits the development of optical anisotropy and mosaic textures are often produced. Chemical adsorption of the pitch molecule to the coke surface should be of major importance. Restriction of anisotropic development is empirically related to the increased viscosity of the slurry, which limits molecular movement. Coking in the microcracks or fissures may also restrict the molecular orientation of the carbonizing species, which can influence the optical anisotropy or layered stacking of carbon planes.

As discussed previously, the surface structure and chemistry of the coke can affect its interaction with the pitch. The carbon surface, even after calcination, always contains some heteroatoms, principally with oxygen functionality, in addition to the graphitic planes. These oxygen functionalities include wettability or increase the surface energy for interaction with pitches. Hence, control of the surface functionalities is very relevant to producing high-quality coke. The selection of feedstocks containing low levels of oxygen, oxygen removal from the feedstock, hydrogen transfer removal of oxygen during the carbonization, and the use of conditions that minimize both oxidation or reductive conditions during calcination can all be used to control the oxygen functional groups at the surface of the calcined coke. Surface treatments performed on carbon fibers may be applicable.

D. Graphitization of Needle Coke and Its Extruded Electrodes

1. Phenomenology of Puffing

The graphitization of needle coke decreases interlayer spacing and increases the size of crytallites, resulting in shrinkage and densification. However, electrodes made from green needle cokes often expand more than expected during graphitization, especially in the radial direction, leaving pores and cracks, thus decreasing bulk density, strength, and conductivities. The irreversible expansion of the electrode after graphitization is known as *puffing* [23–27].

Puffing was a serious problem several years ago when needle cokes with high sulfur content were used as raw materials. Desulfurization of feedstocks or the use of low-sulfur feedstocks solved the problem for Acheson-type graphitization systems. However, in lengthwise graphitization systems, which shorten the graphitization time to save energy consumption, the rapid heating to graphitization temperature has revived puffing as a serious problem, even for needle cokes of rather low sulfur and nitrogen contents.

The diameter of the baked carbon rod artifact generally expands linearly on heating to 1000°C, reflecting the CTE in the c-axis of the aligned needle coke, and then starts to shrink at 1500–1600°C. Above this temperature range, the artifact exhibits a rapid expansion, the extent of which is much larger than that expected from the CTE values. The expansion continues up to 2300°C and then gradually levels off. A small expansion is sometimes observed around 2500°C. The expansions observed in the temperature range 1600–2300°C and at about 2500°C are defined as "dynamic" and "secondary" puffing, respectively. On cooling from 2500°C, the artifact shrinks almost linearly with temperature, rapidly at first until 1000°C, and then gradually. The shrinkage should correspond to the CTE values of the graphitized material. Such a shrinkage–expansion profile is shown in Fig. 26. Such a profile is common to any electrodes; however, the extent of expansion is strongly dependent upon the type of filler coke. The anisotropy of the expansion is clearly observed in the molding of needle cokes of large particle size as shown in Fig. 27.

The experimental results on puffing are summarized as follows:

1. Puffing of a single coke particle is much less than in the extruded rod. This means that coke originating from binder pitch blocks pores in filler cokes to increase puffing [28]. This phenomenon gives a clue to the mechanism of puffing (see the discussion that follows). Figure 28 shows the influence of a pitch impregnation on the extent of puffing.
2. The particle size of the filler coke strongly influences the extent of puffing [29]. Smaller particles lead to less puffing, and filler cokes less than 0.1 mm in the electrode show only a small degree of puffing.
3. Puffing is generally understood to be related to sulfur and nitrogen atoms in the coke. Puffing of petroleum-derived needle coke is believed to be mainly caused by sulfur. In contrast, puffing of coal tar-based needle coke is caused by evolution of nitrogen as well as

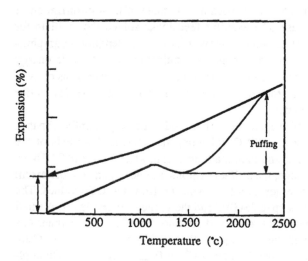

FIG. 26 Profile of shrinkage and expansion curve.

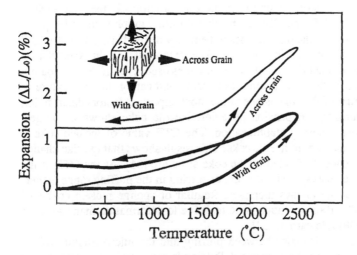

FIG. 27 Influence of measuring direction on the extent of puffing.

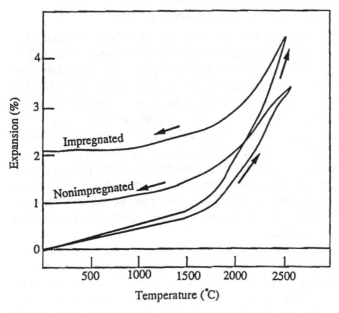

FIG. 28 Influence of pitch impregnation on puffing.

sulfur. Gases originating from these heteroatoms in the coke such as N_2, CS_2, H_2S, and COS are produced during graphitization [30].

4. The extent of the puffing does not entirely depend upon the sulfur and nitrogen contents in the coke. The magnitude of the size change strongly depends upon the kind of cokes used, even if the heteroatom content in the coke is almost the same, indicating that the texture or structure of the coke is one of the most important factors determining the shrinkage–expansion behavior. Needle coke shows the largest expansion during graphitization. The CTE value of mosaic coke is larger than that of needle coke. This result shows that the size change of the coke is affected by the coke structure [31]. As the CTE of the coke decreases, the expansion, occurring in the puffing range, generally but not always increases. Regular or sponge mosaic cokes, of "hard" carbon are much more resistive to deformation than "soft," or graphitizable carbon [24].

5. Several relationships between puffing and the microstructure of the filler coke have been reported. Puffing is characterized by an increase in micropores of diameters in the range 0.1–1 μm [24]. Development of mesopores 1–100 μm in diameter decreases the extent of puffing.

2. Structural Changes in Cokes within Electrodes During Graphitization and the Mechanism of Puffing

The x-ray crystal parameters such as crystallite sizes (L_a, L_c) and interlayer spacing (d) change with heat-treatment temperature (HTT) as shown in Fig. 29. The value of d decreases gradually and slightly from 1200 to 2200°C, whereas crystallite sizes increase rapidly above 1800°C. The changes of the interlayer spacing during graphitization are determined by means of a high-temperature x-ray diffractometer [32].

The results shown in Fig. 30 indicate that the interlayer spacing of a needle coke increases, reflecting thermal expansion; however, it decreases slightly but distinctly at 1700–1800°C where puffing occurs. In contrast, an artificial graphite does not show this behavior. Changes in the crystallite parameter indicate a plastic state of the coke (not yet graphitized), because carbon atoms must move significantly for the completion of graphitization.

Puffing takes place during this plastic state, caused by contaminant gases that push on the pore wall of the graphitizing coke grain as they expand and escape from the middle of the grain as schematically illustrated in Fig. 31. The coke structure is very porous, with a number of micropores (microfissures) spread throughout its whole area, and is

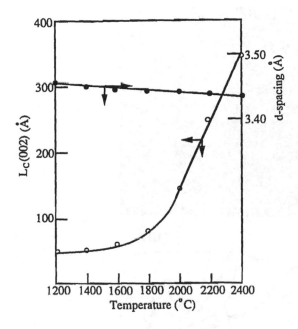

FIG. 29   Changes of crystalline parameters by graphitization.

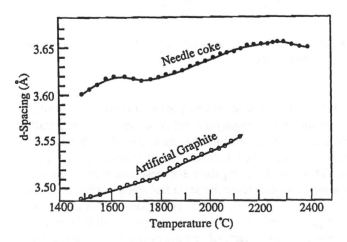

FIG. 30   Changes of *d*-spacing during graphitization.

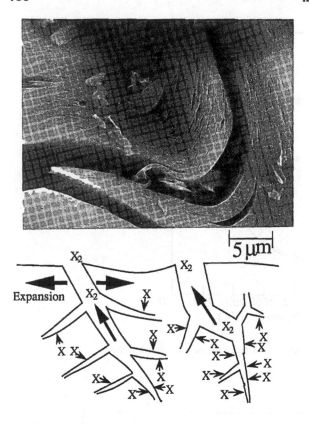

FIG. 31 Model of pore structure in needle coke. X, contaminant atoms; X_2, evolved gas of contaminant.

connected to meso- and then macropores. The heteroatoms such as sulfur and nitrogen in the carbon hexagonal network, which are thermally produced and liberated in the middle of a grain, may flow first into the nearest micropore because the resistance to their transport is at least at that location. The molecules may gather at a mesopore where a number of micropores meet and then leave the grain through a macropore. Such a situation is illustrated in Fig. 31. Puffing takes place when the pressure of the evolved gas exceeds the strength of the pore wall in a manner analogous to how flooding occurs. After a heavy rain fall, the amount of water overwhelms the capacity of river and breaks the bank [33].

If the kinetic momentum of the evolving gases is taken into account, sulfur-containing gases are the main cause of puffing. However the extent of puffing in coal tar needle coke is recognized as being proportional to the nitrogen content, which is modified by a preheat treatment without changing the sulfur content and pore volume [34]. Nitrogen evolution from coal tar needle coke is found to occur, at least partially, in the same temperature range of its puffing while that from petroleum needle coke occurs in a lower temperature range than that of the puffing. Wagner et al., however, rule out the contribution of nitrogen evolution [35]. They propose that the evolution at different temperatures may be ascribed to the different sizes of lamellae and porosity, which are thicker and less porous in coal tar needle coke than in petroleum coke. It has been recognized that the liberation of sulfur at 1000–2000°C enhances graphitization [36,37]. Evolving sulfur may align carbon planes locally in the coke into layered stacks whose layer spacing may approach that of graphite. This is a microscopic indication of puffing caused by atomic movement.

The resistance of the carbon wall to the gas pressure may be influenced by the thickness and the orientation of its microcrystallites. The hexagonal graphitic planes of the pore walls can be moved easily in their c-axis direction by the gaseous pressure. The extent of wall plasticity due to the graphitization process may reduce this resistance.

According to the foregoing considerations, the extent of puffing depends not only on the amount and type of heteroatoms, but also on the rate of gaseous evolution, the detailed structure and graphitization profile, and the heating rate of the graphitization process.

3. Ways to Reduce Puffing

Based on the foregoing mechanistic understanding of puffing in needle coke, some plausible ways to reduce puffing can be proposed. They are classified into four categories [38]:

1. Reducing the contaminant atoms sulfur and nitrogen in the coke feed or coke.
2. Modifying pore distribution in the coke.
3. Changing the heating programs used in calcination and graphitization to control the evolution of contaminant gases.
4. Trapping the heteroatoms or modifying the time of their evolution (accelerating or delaying).

The first category is straightforward, and the approaches for this are the subject of this review. The pore size distribution can be controlled or modified at the calcination as well as at the coking stage. This approach is another subject to be covered here. Slow heating during graphitization can reduce the extent of puffing, although it may not be practical from an economical point of view.

Puffing inhibitors have been applied in electrode manufacturing processes. Iron oxide, which is economical and nontoxic, has been widely used. Many studies have been carried out to elucidate the mechanism of puffing inhibitors. For petroleum needle coke, whose puffing is believed to be mainly caused by sulfur evolution as described previously, puffing is inhibited by adding iron oxide, which reacts with sulfur in the coke [39].

The effectiveness of metals is related to their chemical affinity for sulfur [40]. Metal sulfides are believed to decompose at a higher temperature than that where puffing takes place [41].

For coal tar needle coke, whose puffing is caused by both sulfur and nitrogen, chromium oxide in combination with iron oxide effectively inhibits puffing [42]. Nickel and cobalt oxides can also inhibit puffing [32]. Catalytic activity for graphitization may be related to puffing inhibition.

Some metals may accelerate the carbonization of binder pitch, thus increasing coke yield during baking and leading to a higher density of the electrode. Such increases of coke yield from the binder partially compensate for decreases in density caused by puffing. Iron, nickel, and cobalt oxides increase coke yield, but chromium oxide is inactive. Thus, the acceleration ability to enhance carbonization should be taken also into account when selecting puffing inhibitors [32].

Smaller particle sizes of puffing inhibitors are more effective at increasing the carbonization yield of binder pitch and bulk density, as well as reducing puffing effectively. It is believed that this is due to better dispersion of the fine metal oxides in carbon bodies, and hence closer contact with needle coke, and thus with the sulfur and nitrogen present in the coke [43]. The inhibitors must not be hydrophilic since absorbed water in the electrode may cause expansion, thus reducing its mechanical strength. Because of this, calcium oxide cannot be used as an inhibitor, although it is effective in sulfur capture.

VI. CARBONIZATION IN A TUBE BOMB

Coking in the delayed coker drum leading to needle coke is a target for mechanical study. However, the coker drum is too large to operate, and the reactions in the drum are too heterogeneous for study in the laboratory. The authors have investigated the carbonization process in tube bombs that can control the carbonization pressure and temperature as desired [44,45]. The heating rate can be controlled and can be as rapid as several hundred degrees per minute in sand or melted tin baths. The careful selection of such coking variables produces lump coke in the tube bomb, which is comparable to the needle coke produced from the same

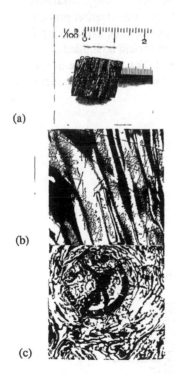

(a)

(b)

(c)

FIG. 32 Lump coke produced in tube bomb and its microscopic appearance. Feedstock, LSVR; carbonization temperature, 500°C; carbonization pressure, 16 kg/cm^2; carbonization time, 4 h; heating rate, 250°C/min. (a) Lump coke; (b) parallel to axis; (c) vertical to axis.

feed stock in a commercial delayed coker in terms of their needle-like appearance, optical texture, and CTE.

Figure 32 illustrates the appearance and photomicrographs of a lump coke produced in a tube bomb. Lump coke is clearly similar to commercial needle coke. Thus, coking in the delayed coker is reproduced and simulated by coking in the tube bomb.

Coking experiments in tube bombs thus provides approaches to (a) help solve the phenomenological and chemical mechanisms leading to needle coke, (b) determine the relation between structure and reactivity of the feed and the properties of the resultant coke, (c) survey for optimum conditions of the respective feeds and their influences on its reactivity, and (d) help study the efficiency and mechanisms of cocarbonization, which is widely applied in commercial coking operations.

Figure 33 illustrates the tube bomb and the carbonization procedure. The whole bomb is immersed in the bath. Because the carbonizing feed is heated externally, the diameter of the bomb should be limited so as not to produce a temperature gradient in the feed. The lump coke is needed for the measurement of the bulk properties of needle coke, so its recovery

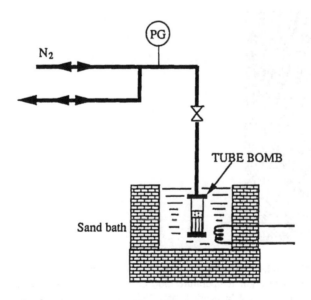

FIG. 33 Tube bomb and heating bath for the carbonization.

requires some care. The authors recommend wrapping the whole feed in aluminum foil or in a can as illustrated in Fig. 34 during carbonization. This ensures that no adhesion of coke to the steel wall occurs. Glass tubing may be used, but heat-transfer problems must be taken into account.

It may be of value to point out differences between coking in tube bombs and coker drums. Table 5 summarizes these differences. The major difference is the feeding procedure. Continuous and batch feedings are for the coker drum and the tube bomb, respectively. No flow of feed, no supply of any volatiles from the lower carbonizing levels, no

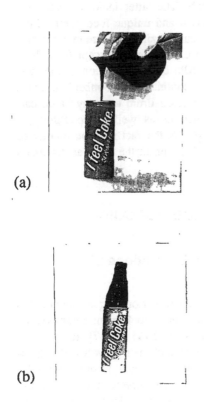

(a)

(b)

FIG. 34   Feed and product coke in a coke can.

TABLE 5 Carbonization Method and Its Conditions

	Delayed coker	Common method	Tube bomb
Heating rate	High	Low	High
Pressure	1–7 kg/cm^{2}G	Atmosphere	Atmosphere—50 kg/cm^{2}G
Recycle oil	Existence	None	None
Carbonization method	Semibatch	Batch	Batch

recycle feed, and no cap of the upper carbonizing levels is expected in batch carbonization in the tube bomb. The latter indicates that each location in the coker drum has its own and unique feed composition as described previously. The variable carbonizing pressure in batch carbonization may compensate to some extent for all except feed flow. Hence, the optimum pressure for needle coke production in the tube bomb may be different from that in continuous carbonization. Thus, one can imagine carbonization in the coker drum as many batch carbonizations of variable feed compositions as well as conditions of temperature and pressure. This image fits the fact that the quality of the coke is variable according to the location in the drum as discussed in Section III.

VII. PHENOMENOLOGY OF NEEDLE COKE FORMATION

A. A Series of Consecutive Observations of Carbonization

The progress of the carbonization process of various organic substances including petroleum residues, coal tar, and model organic compounds has been studied under a microscope since 1965 [46,47]. Brooks and Taylor first observed mesophase spheres and their growth and coalescence into anisotropic coke. Carbonization schemes leading to needle coke, blast furnace coke, and anisotropic liquid crystal pitch for carbon fibers and carbon–carbon composites have been clarified in terms of anisotropic development. However, the route for the use of needle coke as an excellent filler for producing high-quality electrodes has not been satisfactorily clarified, although anisotropic coke is certainly believed to be produced by following similar steps. The reason is that real needle coke, produced in commercial coker drums, has not been completely reproduced in the laboratory, although the tube bomb procedure solved a great deal of the problem, as described in a previous section.

In this section, a study of the carbonization process under an optical microscope is described [48]. A FCC-DO, produced in a Japanese refinery, was carbonized in tube bombs under 16-K g/cm^2 pressure to ascertain if it could provide a lump of typical needle coke. Then, the steps involved in its carbonization were followed by stopping the carbonization at certain periods of time and examining the progress under the microscope. It is important to observe the whole piece of intermediate product because the carbonization never progresses uniformly in whole fractions of FCC-DO. Figure 35 illustrates a series of photomicrographs of the carbonization intermediate products.

Development of optical anisotropy occurs certainly in consecutive steps, which include the formation of mesophase spheres and their growth and coalescence into a bulk mesophase of flow texture, as often

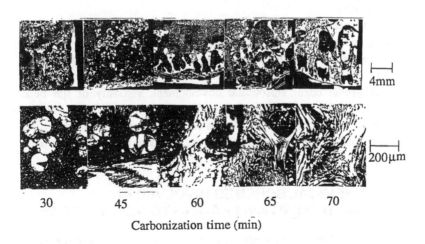

4mm

200μm

30   45   60   65   70

Carbonization time (min)

FIG. 35 Series of montages of the carbons produced from FCC-DO at 500°C in several carbonization times under 16 kg/cm^2 in tube bomb.

reported [49–53]. This carbonization scheme clearly shows the route for the formation of large isochromatic units with the optical anisotropy observed in needle cokes. The extent of anisotropic development is strongly dependent on the carbonization conditions as well as on the feedstocks.

The next stage is unique to the formation of needle coke. The bulk mesophase, produced first in the bottom of the tube and then in the whole region, is rearranged into a flow texture parallel to the tube bomb axis by gas evolution just at the point of solidification of the mesophase into a solid lump of needle coke (Fig. 35). The timing and amount of gas evolved at solidification of the mesophase essentially determine the extent of the orientation in the resultant coke.

B. Carbonization Scheme to Form Needle Coke

Such a series of observation leads to a model of the carbonization scheme as illustrated in Fig. 36, which consists of six major carbonization steps: (1) destructive distillation, (2) sphere formation, (3) growth and coalescence, (4) bulk mesophase laying down parallel to the bottom, (5) growth of bulk mesophase in the whole region, and (6) rearrangement of mesophase planar molecules into uniaxial orientation and, finally, solidification.

The timing of mesophase growth, increase of viscosity and solidification, and gas evolution is strongly influenced by the carbonization temperature and pressure, as well as by the structure and reactivity of the intermediates. Thus, three major factors essentially determine the quality of the needle coke: anisotropic development, viscosity of the bulk mesophase, and gas evolution. Therefore, optimum conditions characteristic of each feedstock, in relation to its reactivity and its structure, must be defined for the best possible needle coke to be produced from it.

VIII. CHEMICAL SCHEMES INVOLVED IN CARBONIZATION LEADING TO NEEDLE COKE

Chemical analyses of the feedstocks and the carbonization intermediates produced in the tube bomb clarify the chemistry of the carbonization reaction leading to needle coke [54–57]. Structural bases for the reactivity of the feed can be also provided.

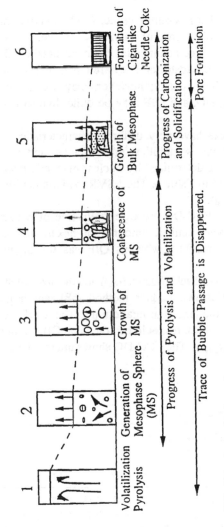

FIG. 36   Carbonization scheme.

A. Chemical Bases of the Feedstocks

Vacuum residue of a low-sulfur crude, fluid catalytic cracking decant oil, and coal tar pitch free of quinoline-insoluble substances (QIF) are discussed here since commercial premium needle cokes have been produced from these feedstocks. They are representative of feeds used in the Japanese coking industry; however, they are not unique. Hence, structures of VR, DO, and QIF may be quite diverse from one source to another.

These feedstocks have very different characteristics, and their elemental compositions, solubilities, aromaticities, and number of alkyl substitution (σ_{al}) as determined by nuclear magnetic resonance (NMR) are all summarized in Table 6. The LSVR exhibits the lowest C/H ratio and aromaticity (f_a).

The LSVR and FCC-DO are essentially free of benzene insolubles (BI) and carry very limited amounts of hexane-insoluble–benzene-soluble (HI–BS) fractions, whereas QIF has the highest content of HI–BS.

The number of aklyl substituents (σ_{al}) and the sizes of structural units, according to the Brown–Ladner method [58], are largest in LSVR, whereas QIF has the largest aromatic units. The alkyl groups in LSVR tend to have long chains. It should be noted that QIF contains a considerable amount of oxygen. The FCC-DO shows intermediate values among the feedstocks.

TABLE 6   Properties of Feedstocks

	Elemental analysis (wt%)								Solubility (wt%)[d]			
	C	H	N	S	O[a]	C/H	f_a[b]	σ_{al}[c]	HS	HI–BS	BI–QS	QI
LSVR	86.2	12.5	0.4	0.2	0.7	0.6	0.18	0.52	92	8	0	0
FCCDO	88.9	9.5	0.1	0.4	0.1	0.8	0.54	0.32	100	0	0	0
QIF	91.6	5.1	1.3	0.7	1.3	1.5	0.88[e]	0.13[e]	48	40	4	0

[a]Difference.
[b]Carbon aromaticity according to Brown–Ladner method.
[c]Degree of substitution of aromatic nucles according to Brown–Ladner method.
[d]HS, hexane soluble; HI–BS, hexane insoluble but benzene soluble; BI–QS, benzene insoluble but quinoline soluble; QI, quinoline insoluble.
[e]HI–BS, hexane insoluble but benzene soluble.

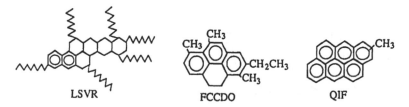

FIG. 37 Structural images of coking feeds. LSVR, low-sulfur crude vacuum residue; FCC-DO, fluidized catalytic cracking decant oil; QIF, coal tar pitch QI free.

Their chemical "average" structures are shown in Fig. 37 and can be derived from the foregoing analytical information as found in the literature [59–61]; however, they are never strictly correct, since these feedstocks consist more or less of complex mixtures of hydrocarbons. Thus, clarification of their structural distribution is required.

Figures 38 and 39 illustrate structural distribution profiles of the feedstocks obtained by thin-layer chromatography (TLC) and gel permeation chromatography (GPC), respectively. The LSVR consists of 50% saturate, 30% aromatic, 4% resin, and 16% asphaltene. The former two fractions are isolated by column chromatography.

The saturate fraction of the feed is essentially paraffinic and gas chromatography (GC) shows that this fraction consists principally of straight chain paraffins, ranging approximately from C_{13} to C_{35}. Table 7 shows some properties of the aromatic fraction in LSVR.

The aromatic fraction of LSVR contains very long alkyl chains, with its f_a (carbon aromaticity) and σ_{al} (number of alkyl substitutions on aromatic rings in a structural unit) very small and large, respectively.

Although the asphaltene fraction is very difficult to characterize, it contains some condensed aromatic rings that are connected by methylene bonds into polymeric chains. The fraction is believed to contain a number of long alkyl substitutes and to form micelles, dispersed in the paraffinic and aromatic matrix as proposed by Yen et al. [62]. Sulfur is concentrated in the asphaltene. The GPC indicates its widely dispersed molecular weight, ranging from 800 to 7000. Such dispersed compositions may strongly influence the carbonization properties of the whole vacuum residue through their mutual interactions.

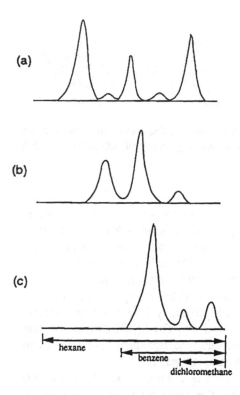

FIG. 38 TLC–FID chromatographs of feedstocks. a, LSVR; b, FCC-DO; c, QIF.

Both FCC-DO and QIF have much narrower molecular distributions as indicated by GPC (800–2500), although the latter feed consists of BS (96%) and BS-QS (4%) fractions.

The former feed consists essentially of saturate (28%) and aromatic (70%) fractions as indicated by TLC. The GC shows that this saturate fraction also consists of straight paraffins ranging from C_{20} to C_{30}. The aromatic fraction appears to carry more or less alkyl chains of variable length as shown in Table 7. The f_a of the fraction is 0.74, much higher than that of the same fraction in LSVR. The subfraction of more aromatic rings with fewer and shorter alkyl groups and fewer naphthenic groups is less soluble or absorbed more strongly on alumina to be eluted later.

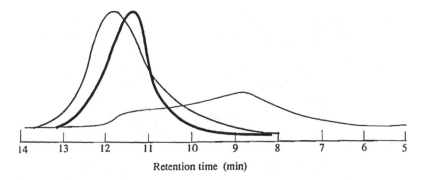

Retention time (min)

FIG. 39 Gel permeation chromatographs of feedstocks. —, LSVR; —, FCC-DO; ▬, QIF.

Detailed structures of respective and aromatic subfractions are illustrated in Fig. 40, based on FD-MS and ^{13}C-NMR in addition to the previous analyses [63].

QIF consists of aromatic, resin, asphaltene, and BI–QS fractions. They are all essentially polycondensed aromatic hydrocarbons of variable sizes (3–10 rings) with the least alkyl and naphthenic groups, f_a being around 0.9 regardless of the fractions. It should be noted that this feedstock carries a considerable amount of phenolic groups as indicated by Fourier transform infrared spectroscopy (FT-IR). The oxygen content is also a major influence on the solubility; higher oxygen content leads to lower solubility. The aromatic rings are bonded through aryl–aryl (major) and methylene (minor) bonds, defining molecular weights and also solubilities of the fractions.

TABLE 7 Properties of Aromatic Fractions of Feedstocks

	Elemental analyses (wt%)					^{1}H–NMR[b]					
	C	H	N	O[a]	C/H	H_a	H_α	H_β	H_γ	$f_a{}^a$	$\sigma_{al}{}^a$
LSVR	87.9	10.3	0.6	1.2	0.7	10	17	57	16	0.40	0.51
FCCDO	91.8	7.1	0.1	1.0	1.1	42	41	13	4	0.74	0.34

[a]See Table 6.
[b]H_a; 10–6 ppm, H_α; 4–2 ppm, H_β; 2–1.1 ppm, H_γ; 1.1–0.5 ppm.

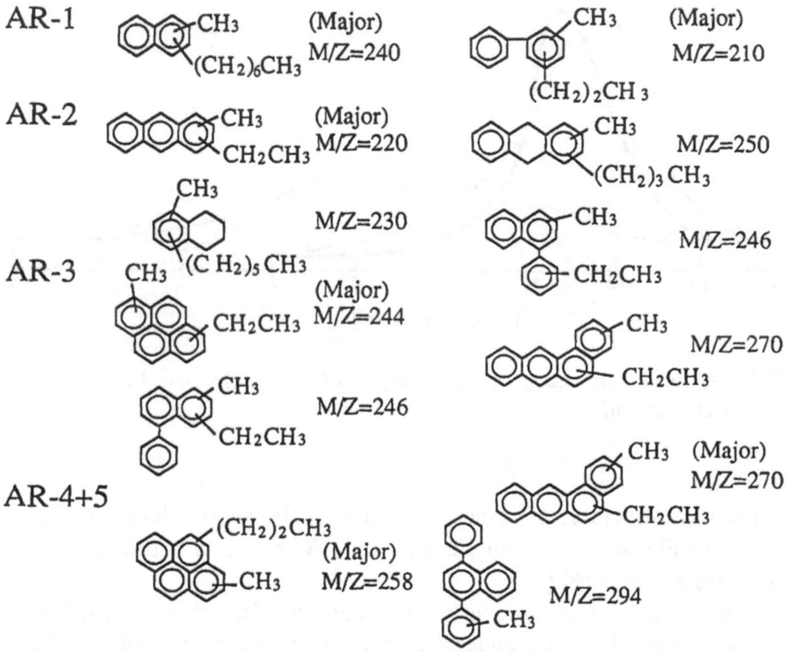

FIG. 40 Detailed model structure of aromatic fractions in a representative FCC-DO.

B. Compositional Changes of the Feeds during Carbonization

Figure 41 shows the amount of the intermediate and its solubility during the carbonization of the feedstocks at 480°C under 16 kg/cm^2. The LSVR, initially completely soluble in hexane, decreased to 49% of the starting quantity within 0.5 h. The intermediate remaining in the tube at this time consists of 26% HS, <7% HI–BS, 10% BI–QS, and 6% QI on a feedstock basis. A longer carbonization time further decreases the amount remaining in the tube and increases the content of QI. The carbonization appears to be completed by 2 h because the weight of QI at 2 h is the same as that at 6 h, although 27% of QS still remains in the tube. The period of solidification, which determines the final anisotropic

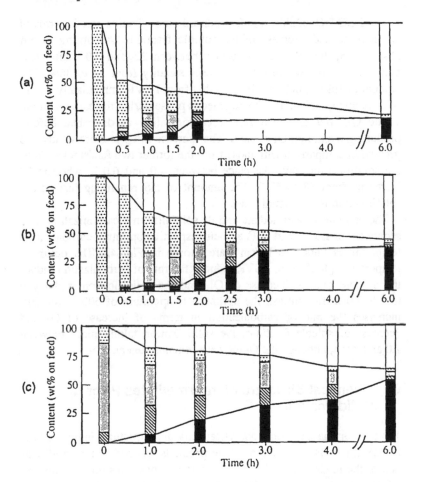

FIG. 41 Solubility of intermediates produced at 480°C under 16 kg/cm^2. Feedstock: (a) LSVR; (b) FCC-DO; (c) QIF. □ , effluent; ▦ , HS; ▨, HI–BS; ◩, BI–QS; ■ , QI.

texture, may be located between 1.5 and 2.0 h after the carbonization started, while QI increased from 7 to 17%, dominating the intermediate. The increase of QI for the initial 1.5 h is rather slow and becomes rapid on solidification. It should be noted that 20% of HS remains just before solidification. At this stage, HS and QI may form separate phases.

The FCC-DO exhibits a much reduced carbonization rate in terms of QI increase and decrease of the soluble fractions. According to our observation, the solidification occurs between 2.5 and 3.0 h after carbonization starts, whereas QI increases steadily from 22 to 36%. The amount of HS is much smaller than that from LSVR immediately before solidification, indicating a homogeneous phase during the carbonization.

The QIF reveals much slower carbonization as described before, producing maximum QI at 6.0 h. The increase in QI is more steady and gradual in comparison with those from the former two feedstocks. Thus, the solidification may take place between 4.0 and 6.0 h, whereas QI increases from 38 to 67%. The amount of HS remaining just before solidification is very small (2%).

The amount of intermediate and its solubility in a carbonization time sequence for LSVR and QIF, at their respective optimum carbonization temperatures under 16 kg/cm^2, are shown in Fig. 42. The optimum temperature (460°C) for LSVR reduces the rate of carbonization inducing a more gradual increase of QI, and the solidification is delayed to 2.5–3.0 h. In contrast, the optimum temperature of 500°C for QIF increases the rate of carbonization in terms of increase of QI and decrease of soluble fractions, the solidification taking place between 2.0 and 3.0 h, although the QI increase remains steady.

C. Chemical Structure of Intermediates Prior to Solidification

Table 8 summarizes some analytical data and structural indices of the HI–BS fractions in the intermediate products just before the solidification of the feedstocks at 480°C. The fraction may constitute the matrix, influencing the viscosity of the carbonizing system. At the same time, structural characteristics may indicate the chemical changes during carbonization. The carbon aromaticity (f_a) of the intermediate from QIF (QIF-I) is largest, whereas that of LSVR-I is smallest. In contrast, the number of substituted groups per unit structure of QIF-I is smallest, and that of LSVR-I is largest. The FCC-DO-I shows intermediate values in terms of these structural indices. Thus, the intermediates of the feedstock strongly reflect their initial structure.

Figure 43 illustrates TLC compositions of BS fractions in LSVR-I and FCC-DO-I, respectively. The fraction from LSVR-I consists of 9% saturate, 67% aromatic, 9% resin, and 15% asphaltene. The GC shows

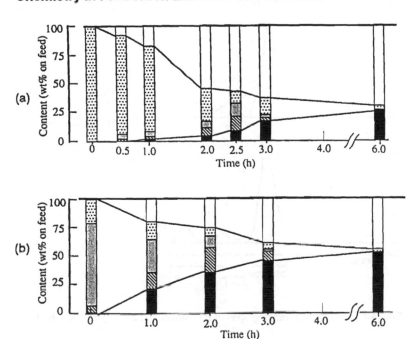

FIG. 42 Solubility of intermediates produced at the respective optimum temperature under 16 kg/cm^2. Feedstock: (a) LSVR; (b) QIF. □ , effluent; ▦ , HS; ▨ HI–BS; ◩ , BI–QS; ■ , QI.

TABLE 8 Properties of HI–BS Fraction in the Intermediates Produced Just Before the Solidification at 480°C Under 16 kg/cm^2

	Carbonization time (min)	Elemental analyses (wt. %)						
		C	H	N	O^a	f_a^a	σ^a	C/H
LSVR-I	90	91.2	5.8	1.0	2.0	0.72	0.37	1.3
FCC-DO-I	150	92.6	5.8	0.1	1.5	0.89	0.17	1.3
QIF-I	240	92.8	5.0	1.0	1.2	0.96	0.08	1.5
LSVR-I-A^b	90	91.6	7.0	0.3	1.1	0.74	0.30	1.1
FCC-DO-I-A^b	150	92.8	6.0	0.1	1.2	0.88	0.19	1.3

^{a}See Table 8-1.
^{b}Aromatic fraction separated by column chromatograph.

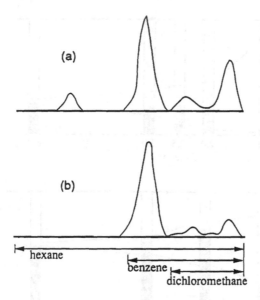

FIG. 43 TLC–FID chromatographs of feedstocks: a, LSVR-I; b, FCC-DO-I.

that the saturate consists of straight paraffins, ranging from C_{13} to C_{30}. It should be noted that these straight paraffins remain in the intermediate, although their content is much lower than that in the starting feedstock. The aromatic hydrocarbons still carry a considerable amount of long side chains. The fraction FCC-DO-I essentially consists of aromatic hydrocarbons, comprising mainly short alkyl chains. The saturate in FCC-DO disappears completely at the early stage of carbonization, suggesting its high reactivity for pyrolysis.

Figure 44 shows the GPC of HI–BS fractions of the intermediates. Molecules in the LSVR-I fraction are the largest and most dispersed in spite of the short carbonization time and its complete solubility in benzene, whereas those of FCC-DO and QIF are smaller and their distributions are narrower. Such structural characteristics may suggest structural continuity of the carbonization intermediates from FCCD and QIF. Even though that from LSVR still has the form of straight paraffins and long side chains, the heavy asphaltene and mesophase fractions seem anxious to precipitate and to be carbonized separately.

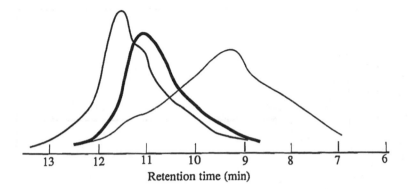

Retention time (min)

FIG. 44 Gel permeation chromatographs of HI–BS fractions of the intermediates. —, LSVR, — , FCC-DO; — , OIF.

D. Gas Evolution and Its Composition

Figure 45 illustrates profiles of gas evolution during carbonization of
the respective feedstocks at 480°C and 16 kg/cm^2. The regions of
solidification are also indicated in the figure. The amount of gas
evolved from LSVR increases sharply to a maximum at 40 min,
and then decreases to zero at 150 min. The rate of evolution at
1.5–2.0 h, which is the region of solidification, decreases from 35 to
8 ml/10 min.

The quantities of gas evolved from FCC-DO increase to reach
a maximum rate at 50 min, which is lower than that from LSVR,
and then decrease gradually. The rate of evolution in the solidification range decreased from 14 to 6 ml/10 min. In marked contrast to these feedstocks, the amount of gas evolved from QIF is
very small, and the evolution rate at solidification is as little as
2 ml/10 min.

The profiles of gas evolution from the three feedstocks at several
carbonization temperatures are shown in Fig. 46. The evolution of gas
from LSVR at 460°C is more delayed than that at 480°C. The rate of gas
evolution at solidification decreases from 13 to 3 ml/10 min, which is
similar to that observed in FCC-DO at 480°C.

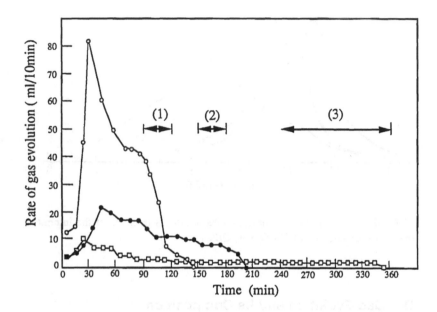

FIG. 45 Profile of gas evolution during carbonization at 480°C under 16 kg/cm². O, LSVR; ●, FCC-DO; □, QIF. Region of solidification for (1) LSVR, (2) FCC-DO, (3) QIF.

The amount of gas evolution from QIF at 500°C, its optimum carbonization temperature, is a little more upon solidification than at 480°C, although at 500°C it is still <6 ml/10 min, which is much smaller than that from FCC-DO at 480°C.

The carbonization pressure hardly influences the rate of gas evolution; however, it varies the solidification time so that the amount of the gas evolution at solidification is affected by pressure.

The gases evolved from FCC-DO and QIF at the solidification stage consists almost exclusively of C_1 and C_2 paraffins, although those of LSVR contain a considerable amount of C_3–C_5 in addition to C_1 and C_2 paraffins. The composition and quantity of evolved gas should reflect the structure, especially the alkyl side chains, of the intermediates as already described.

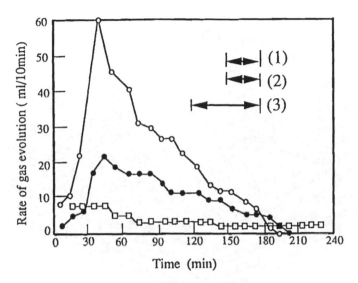

FIG. 46 Profile of gas evolution during carbonization at optimum temperature under 16 kg/cm^2. O, LSVR; ●, FCC-DO; □, QIF. Region of solidification for (1) LSVR, (2) FCC-DO, (3) QIF.

E. Mesophase; Intermediate Liquid Crystal State of Carbonization

The mesophase, carbonaceous liquid crystal is the intermediate that essentially determines the structure and properties of the resultant needle coke, through its chemical and physical transformation of solidification. Hence, the structural understanding of mesophase is a key to finding ways to control carbonization. The mesophase produced in the carbonization and separated from the matrix is hardly soluble in any solvent, and its analysis is not easy. Recent preparation of mesophase pitch-based carbon fiber provided a highly soluble anisotropic pitch, and its analytical study is far more advanced than for other systems.

Figure 47 illustrates model molecules in the mesophase spheres, which are produced at an early stage of the carbonization [64,65]. Their analysis is performed after solubilization by reduction and reductive alkylation. One can see from the figure the following structural characteristics of mesophase molecules.

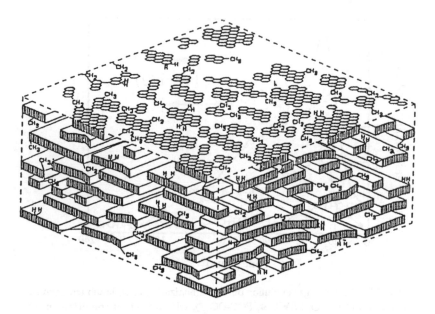

FIG. 47 Model for mesophase liquid crystal.

1. The mesophase consists of a series of molecules whose molecular weight is distributed from 400 to 10,000.
2. Molecules are oligomeric, and they have a skeleton of aromatic nuclei that are connected with aryl–aryl and methylene linkages. The size of the nuclei varies between 2 and 10 rings.
3. Each nucleus can carry a restricted number of methyl and naphthenic groups.

From such molecular model, the mesophase can be a liquid crystal because of the complex mixture of aromatic molecules. Larger molecules are stacked to exhibit diskotic anisotropy, and smaller molecules may cause the thermotropic fluidity, being located between the diskotic layers. Such an understanding leads to the phase diagrams shown in Fig. 48 [66,67].

The properties of the mesophase are thus governed by the temperature and pressure according to its thermotropic nature and the composition of constituent molecules according to the nature of the lyotropic or cooperative liquid crystal and the structure of each constituent molecule. Hence,

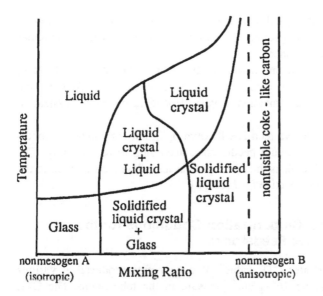

FIG. 48 Phase diagram of a liquid crystal mixture of isotropic nonmesogen (A) and anisotropic nonmesogen (B).

its properties are controllable through the carbonization conditions, composition adjustment, and structural control. It has been widely documented that naphthenic and alkyl groups in the constituent molecules assist the fusibility or lower the temperature range for the liquid crystal phase. The chemical basis of cocarbonization may be clarified. Thus, the structural understanding of mesophase at its molecular level provides information on ways to chemically control its properties.

IX. THE OPTIMIZATION PRINCIPLE FOR NEEDLE COKE PRODUCTION

Needle coke is produced via several carbonization steps including mesophase development and uniaxial orientation through consecutive steps during chemical and physical conversion and realignment of mesophase planar components. Hence, the reactivities of the organic feed substances for dehydrogenation, dealkylation, aromatization, aromatic condensation, and gas evolution, proceeding through the mesophase liquid crystal state and finally to the solid coke, should be suitably balanced to provide a resultant needle coke with a particular structure and appearance.

The reactivities of the organic components vary from feed to feed, reflecting its chemical composition as well as structure. At the same time the rate of carbonization is very dependent on the carbonization conditions.

Thus, the respective feed has its own optimum carbonization conditions to provide premium needle coke according to its reactivity. The optimization principle is thus discussed as an intimate connection to the reactivity of the respective feed [68,69].

A. Optimum Carbonization Conditions for the Respective Feedstocks

Figure 49 illustrates CTE values of coke lumps produced at various temperatures under 16 kg/cm^2 pressure in the tube bomb. The CTE values of the cokes vary drastically depending upon the carbonization temperature. The optimum temperature can be defined as the temperature that provides the lowest CTE of the coke.

Each feedstock has its own optimum carbonization temperature. The optimum temperatures of the QIF, FCC-DO, and LSVR are found to be 500, 480, and 460°C, respectively. It should be noted that the lowest CTE values, 0.3×10^{-6}/°C, obtained by carbonization of FCC-DO at the optimum temperature of 480°C are beyond that of commercial cokes at present.

The times for carbonization completion are dependent upon both the temperature and the feedstock, which reflect the reactivity of each of the different feedstocks. The LSVR and FCC-DO take 2 and 3 h, respectively, for complete carbonization under 16 kg/cm^2 at 480°C. For QIF it takes more than 10 h to complete the carbonization at the same temperature under 16 kg/cm^2, whereas it takes only 1 h at 550°C, which indicates a fairly large activation energy.

Table 9 summarizes the CTE values of coke lumps produced under a series of pressures from LSVR and QIF. Higher carbonization pressures tend to provide lower CTE values of coke lumps from both feeds at 500°C. It is often reported that mesophase development is better under higher pressure. However, QIF shows the optimum carbonization pressure, giving the lowest CTE in carbonization at 550 and 470°C. At

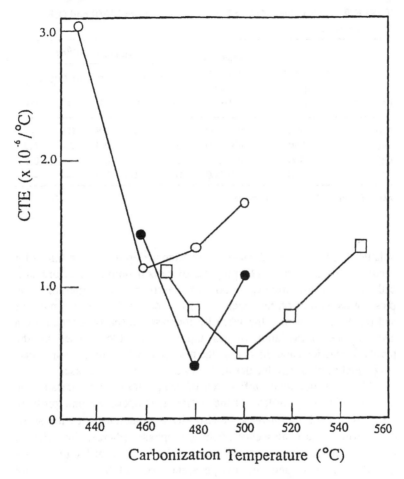

FIG. 49   CTE values of coke lumps produced at various temperatures under 16 kg/cm^2 from feedstocks. O, LSVR; ●, FCC-DO; ◻, QIF.

550°C, 16 kg/cm^2 is the optimum pressure, whereas 1 kg/cm^2 is optimum at 470°C.

Such optimum conditions for needle coke production can be argued from the mechanistic views of carbonization described in Section VIII. The cooperation of temperature and pressure, both of which influence the carbonization reaction through the extents of thermal activation and

TABLE 9 CTE Values of Coke Lumps Produced from LSVR and QIF Under Pressure

| | | Measured | CTE values ($\times 10^{-6}$/°C) | | | | |
| | Carbonization | temperature | Carbonization pressure (kg/cm^2) | | | | |
Feedstock	temperature (°C)	range (RT–°C)	1	8	16	26	41
LSVR	500	RT–100	—	1.2	0.8	0.8	—
QIF	550	RT–500	2.3	1.4	0.3	2.0	1.9
QIF	500	RT–500	1.5	—	0.5	—	0.1
QIF	470	RT–500	0.5	1.0	1.2	—	—

Abbreviation: RT, Room temperature.

volatiles held in the liquid phase, may define the optimum conditions for needle coke production. The progress of carbonization varies dramatically and depends strongly on both the carbonization temperature and pressure as well as on the reactivity of the feedstock. These conditions and reactivity influence the viscosity increase caused by condensation and the rate of evolution of gas and other volatiles caused by the pyrolytic reactions and thus affect the mesophase development and rearrangement of planar molecules in the resultant bulk mesophase.

Table 10 summarizes influences of temperatures and pressure on the carbonization profiles. At the optimum temperature and pressure for a given feed, the increase of viscosity is moderate, allowing sufficient growth and coalescence of the mesophase spheres. This gives a bulk mesophase of low viscosity because the increase of the quinoline-insoluble fraction and the disappearance of soluble fractions are both moderate. At the same time, the amount of evolved gas at solidification is sufficient to rearrange the molecules of the bulk mesophase into a needlelike orientation. This happens because carbonization, under appropriate conditions, is completed within an adequate period of time.

At too high a carbonization temperature, all the physical and chemical changes take place very rapidly. There is little time available to allow the growth of the isochromatic area. The development of the bulk mesophase and the maximum in the evolution of turbulent gas tend to overlap, so that the mesogen molecules are arranged in a random fashion. There is

TABLE 10 Influence of Carbonization Condition on Carbonization Profile and Coke Quality

Carbonization condition		Carbonization profile		Quality of resultant coke		
Temperature	Pressure	Growth of mesophase	Gas evolution at resolidification	Appearance	Texture	CTE
Lower	Higher	Excellent	Too little	Lump	Domain	Higher
Optimum	Optimum	Excellent	Optimum	Lump	Uniaxial flow	Lower
Higher	Lower	Fair	Too much	Flaky	Elongated mosaic	Higher

little chance of forming needle coke with this type of development. Flaky cokes with mosaic textures are usually the result.

When the temperature is too low, the carbonization process is very slow. As a result, carbonization of the components and hence cocarbonization may not occur concurrently because of their different reactivities (and activation energies) and lower mutual solubilities at the lower temperature. Gas evolution is slow, and there is not enough flow at the solidification stage of the bulk mesophase to force the uniaxial arrangement of the mesophase molecules. Again, needle coke is not likely to be produced.

Higher than optimum pressure delays solidification because the lighter fractions in the feed do not vaporize as quickly as under optimum pressure and tend to dissolve the heavier fractions. This favors the growth of the mesophase. However, because pressure does not retard gas evolution to a great extent (due to unimolecular pyrolysis), evolution at higher pressure tends to occur well before the solidification temperature. Therefore, no rearrangement of the bulk mesophase can be achieved. Carbonization pressures of more than 40 kg/cm^2 deteriorate the uniaxial arrangement of the mesophase from QIF (which has the least amount of evolved gas) because of high aromaticity, particularly at lower temperatures, even though the mesophase grows well into a domain texture.

B. Reactivity of Feedstock

The reactivities of the feedstocks for needle coke are as important for the rate of aromatic condensation leading to the mesophase liquid crystals of large anisotropic units as for the rate and amount of gas evolution. Both are driving forces for unaxial rearrangement. Such reactivities originate from the structure of feedstocks. It has been suggested that these reactions are not always unimolecular; hence, the mutual interaction of the components in the feed should play some important roles, as described in detail in the next section.

The optimum conditions of the present three feedstocks are summarized in Table 11. The reaction can be conventionally described by the time of carbonization completion or solidification into green coke.

As can be seen, the most reactive LSVR requires the lowest carbonization temperature at higher pressure to produce good-quality needle coke of excellent texture by delaying the carbonization. The FCC-DO of moderate reactivity, which appears to be the best feed to produce needle coke, achieves its optimum carbonization conditions at a medium temperature of 480°C under moderate pressure to balance the carbonization progress for solidification and gas evolution.

The least reactive feed is QIF, which requires the highest temperature of 500°C for optimum carbonization. Influences of pressure are very sensitive and are very dependent on the carbonization temperature. The optimum pressure is moderate at 550°C for the gas evolution during the solidification stage. However, a higher pressure is better at 500°C, because solidification and gas evolution are better balanced. Under the conditions, even lower gas evolution may allow better rearrangement because of the lower viscosity of the intermediates. At the lowest

TABLE 11 Relation Between Reactivity of Feedstock and Optimum Conditions

	Feedstock		Optimum conditions	
	Reactivity	Gas evolution at resolidification	Temperature	Pressure
LSVR	High	Large	Low	High
FCC-DO	Moderate	Moderate	Medium	Moderate
QIF	Low	Low	High	Moderate

temperature of 470°C, it is best to accelerate the carbonization under lower pressure.

The feed reactivity, as it relates to carbonization, can be qualitatively correlated principally to the structure of the feed. Coal tar pitch of a highly aromatic nature and with the smallest number of substituents is the most stable feed and has the lowest gas evolution. The LSVR is very reactive and has a high gas evolution because of many alkyl groups in a long chain. The FCC-DO is of moderate aromaticity and contains a moderate amount of substituents. It therefore exhibits moderate reactivity and gas evolution. As a result, FCC-DO is the best feed to produce the best needle coke under its optimum conditions.

The roles of noncarbonizing fractions in the feed are discussed in later sections.

X. COCARBONIZATION PROCEDURES TO PRODUCE BETTER NEEDLE COKE

A. Prospect of Cocarbonization

According to the mechanism of needle coke formation described in the preceding sections, the rate of carbonization that influences the formation of mesogens in the early stage, viscosity increase, and gas evolution at the solidification stage are the major factors that define the coke quality. Hence, control of these factors, not only by the carbonization conditions but also by reactivity modification, is a key approach for producing higher-quality needle coke from a given feedstock through simpler procedures. There are two major approaches to modifying the feedstock.

1. Modification of composition
2. Modification of structure

The first approach is discussed in this section. First of all, as briefly indicated in previous sections, carbonization reactions are often bimolecular the reaction of a constituent being very much influenced by the coexisting molecules. Second, the liquid phase and the viscosity of the complex mixtures of hydrocarbons are always defined by the constituent species. Third, the gas evolved from only a part of the constituents is enough for the alignment of whole species during solidification when the evolution is timed properly. Hence, the composition of the feedstock is a very important factor for producing high-quality needle coke, and the removal or addition of a particular fraction from or to the feedstock can control the carbonization properties.

The cocarbonization procedure adds or blends some additives to the carbonization feedstock in order to control its carbonization properties. This has been practical for some time in the coking industry for blast furnace and needle cokes.

The mechanistic understanding of cocarbonization in terms of aniso-tropic development, graphitization, and density of the cokes has been discussed in detail by Marsh and associates [70–71] and Mochida and co-workers [72–74]. Another application of cocarbonization for pitch based carbon fibers has been proposed successfully by Mochida and co-workers [75–77].

Control of anisotropic development through cocarbonization is pro-posed to follow three mechanisms:

1. Liquid–liquid blending to control the viscosity change during car-bonization.
2. Dissolution and solvation of solid products of the carbonization to control the carbonization phases.
3. Solvolytic reactions such as hydrogen transfer, trans-alkylation, and Diels–Alder addition among the constituent molecules to control reactions of the intermediate species.

Cocarbonization to control the particular scheme for the production of needle coke requires further consideration of the uniaxial rearrangement. The study of carbonization in the tube bomb clarifies the cocarbonization scheme for producing better needle cokes.

B. Cocarbonization of FCC-DO and VR; Features and Mechanism

Delayed coking in the petroleum industry has applied the cocarboniza-tion of FCC-DO and LSVR or tar produced by the pyrolysis of vacuum gas–oil [78]. Wider varieties of cocarbonization blending are surveyed for future application. The cocarbonization of FCC-DO and LSVR studied using the tube bomb will now be reviewed [79,80].

The coke yields of the FCC-DO and LSVR mixtures, which are much the same under both conditions of 460 and 480°C at 8 kg/cm^2, coincide approximately to the calculated average of those of FCC-DO and LSVR.

The CTE values of the calcined lump cokes are improved by cocar-bonization as shown in Table 12. The addition of LSVR to FCC-DO

TABLE 12   CTE and Anisotropic Development of Lump Cokes

		FCC-DO	7 : FCC-DO 3 : LSVR	5 : FCC-DO 5 : LSVR	3 : FCC-DO 7 : LSVR	LSVR
	l_{av} (μm)[a]	19.6	20.0	18.4	17.8	16.2
480°C	f_{av} (μm)[a]	16.0	17.3	16.8	16.5	15.0
	CTE × 10^{-6}/°C	0.77	0.10	0.23	0.39	1.30
	l_{av} (μm)	19.0	20.0	19.2	18.2	17.5
460°C	f_{av} (μm)	16.0	16.0	16.6	16.2	15.4
	CTE × 10^{-6}/°C	0.78	0.90	0.36	0.75	1.20

[a]See Fig. 19.

certainly improves (i.e., lowers) the CTE. In terms of CTE the best cokes are produced at the mixing ratios of 7/3 and 5/5, respectively, at 480 and 460°C.

The values of f_a (Section IV) of the feeds are again correlated to their CTE values. Hence, the advantage of cocarbonizations in terms of CTE of the resultant cokes can be discussed in connection with the uniaxial arrangement of anisotropic flow texture.

The completion of the carbonization is defined by the amounts of quinoline-insoluble mesophase and the soluble fraction. The times required for the completion depend on the carbonization temperature and the feedstocks. The FCC-DO is much more stable that LSVR, requiring 2.5–3.0 h for the completion of its carbonization at 480°C, whereas LSVR requires 1.5–2.0 h as described earlier. Addition of LSVR shortens the time, depending on the amount of LSVR in the cocarbonization feed.

The carbonization is very slow at 460°C compared to that at 480°C. The FCC-DO and LSVR take 12.0–12.5 h and 5.0–5.5 h, respectively, for completion of their carbonization. Again, the addition of LSVR accelerates the carbonization to shorten the time in comparison to that of FCC-DO alone at the lower temperatures.

The shorter time for the completion of the cocarbonization (in comparison with that of FCC-DO) can be ascribed to hydrogen transfer from the naphthenic units to the reactive species of the cocarbonizing substances. As shown in Table 6, LSVR carries a considerable amount of asphaltene, which is the major origin for the coke and has a high reactivity because of the large molecular weight of the aromatic nuclei, many alkyl side chains, and micelle agglomeration. The paraffinic fraction of another major component in LSVR is a poor solvent for the asphaltene fraction, especially after the asphaltene molecules lose their alkyl side chains during the carbonization. Hence, mesophase of low viscosity is difficult to produce from LSVR alone, and smaller anisotropic units tend to be introduced. In contrast, rather stable aromatic species in FCC-DO provide bulk mesophase of low viscosity. Its naphthenic structure helps such a mesophase preparation. Hence, the cocarbonization of FCC-DO and LSVR controls the high reactivity and low solubility of the asphaltene fraction in LSVR to provide better mesophase from their mixture.

Figure 50 illustrates the amount of remaining carbonaceous substance in the tube bomb and its solubility during the carbonization of the feedstocks at 480°C under 8 kg/cm^2. The FCC-DO is initially all soluble in benzene, but solubility decreases to 56% of the starting amount after 1.0 h of heating. The remaining substances in the tube at this time consist of 7% HS, 36% HI–BS, 12% BI–QS, and 1% QI on the feedstock basis. Longer carbonization time further decreases the remaining amount of soluble fractions and increases QI. The carbonization appears to be complete at about 3.0 h because the weight of QI at this time was the same as that at 12.0 h. The period of solidification that may determine the final features of the anisotropic texture may be between 2.5 and 3.0 h after the start of carbonization, whereas QI increases from 8 to 38% (remaining carbonaceous substance basis, 92%) dominating the intermediate.

Addition of LSVR at a ratio of 7/3 (FCC-DO/LSVR) accelerates the carbonization in terms of QI increase as well as the decrease of soluble fractions. The carbonization follows a typical scheme for the formation of a well-developed bulk mesophase, similar to that of FCC-DO, except for the formation of bottom mosaic coke at a very initial stage, while a large amount of the HS fraction still remains in the bomb. Anisotropic spheres appear after the precipitation of bottom coke (mosaic texture) and bulk mesophase is produced on the bottom mosaic coke after 2.0 h. Uniaxially oriented flow textures are observed in the middle portion of the coke lump produced at 2.5 h. The solidification takes place between 2.5 and 3.0 h after the carbonization has started, whereas QI increases from 13 to 34%. It should be noted that 9% of HS still remains in the intermediate just before the solidification.

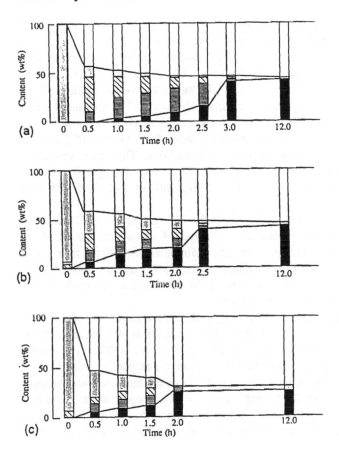

FIG. 50 Solubility of intermediates produced at 480°C under 8 kg/cm^2. Feedstock: (a) FCC-DO; (b) 7/3(FCC-DO/LSVR) mixture; (c) LSVR. □ , effluent; ▨ , HS; ▨ ; HI–BS; ▤ , BI–QS, ■ , QI.

The LSVR alone exhibits a very rapid carbonization, as described in Section VIII. Bottom mosaic coke is produced in a manner very similar to that observed with the mixture. Many spheres were produced; however, their growth was restricted to form coarse mosaic textures in the major areas. The solidification of the mesophase took place between 1.5 and 2.0 h, whereas QI increased from 11 to 23%. Just before the solidification, 15% of HS still remained.

Table 13 summarizes some analytical data of HS fractions in the intermediate products just before the respective solidification of the feedstocks at 480 and 460°C. The fraction in the intermediate from LSVR at 480°C consists of 9% saturate, 87% aromatic, 1% resin, and 3% asphaltene. The GC shows that the saturates consists principally of straight chain paraffins, ranging from C_{11} to C_{24}. The aromatic fraction consists of compounds of 1–2 rings with long alkyl side chains.

The fraction of the intermediate from FCC-DO at 480°C consists almost exclusively of aromatic hydrocarbons. The small amount of saturates in this fraction disappears completely during carbonization. The aromatic fraction consists of compounds of 1–2 rings with short alkyl side chains.

The fraction of intermediate from the 7/3 mixture at 480°C consists of 2% saturate, 93% aromatic, 2% resin, and 3% asphaltene. The GC shows that the saturate consists of straight chain paraffins, ranging from C_{16} to C_{24}. The fraction of the intermediate from a 7/3 mixture at 460°C does not include any saturate components. In contrast, the fraction from a 5/5 mixture carries 16% of the saturate.

Table 14 summarizes some analytical data and structural indices of the HI–BS fractions in the intermediate products of the feedstocks just before the respective solidification at 480 and 460°C. The carbon aromaticities (f_a) of the intermediates from the respective feedstocks are almost the same (around 0.9), regardless of the feedstock and temperature. Even though the H_β value (content of hydrogen, its chemical shift distributing 2 ~ 1.1 ppm, is largest for the intermediate from LSVR; the R_{nus} (fraction of naphthenic rings in a structural unit) value is largest with

TABLE 13 Properties of HS Fraction of Intermediates Just Before Solidification

Feedstock	Temperature (°C)	Time (h)	TLC-FID				GC Paraffin
			Saturate	Aromatic	Resin	Asphaltene	
FCC-DO	480	2.5	0	96	2	2	—
7/3	480	2.0	2	93	2	3	C_{16}–C_{24}
LSVR	480	1.5	9	87	1	3	C_{11}–C_{24}
7/3	460	11.5	0	98	1	1	—
5/5	460	10.5	16	61	21	2	C_{17}–C_{24}

TABLE 14 Properties of HI–BS Fraction of Intermediate Just Before Solidification

Temperature (°C)	Feedstock	Time (h)	CH	^{1}H-NMR[a]				$f_a{}^{b}$	$R_{nus}{}^{c}$
				H_a	$H\alpha$	$H\beta$	$H\gamma$		
480	FCC-DO	2.5	1.5	55	26	14	5	0.86	2.77
480	7/3	2.0	1.5	64	20	13	3	0.89	0.98
480	LSVR	1.5	1.4	57	19	21	3	0.85	0.53
460	7/3	11.5	1.6	71	9	18	2	0.91	0.36
460	5/5	10.5	1.5	69	14	13	4	0.90	0.17

[a]See Table 7.
[b]See Table 6.
[c]Fraction of naphthenic ring in structural unit according to Brown–Ladner method.

the intermediate from FCC-DO, whereas the value for the intermediate from LSVR was the smallest.

The viscosity and its change should reflect the structure and compositions of the mesophase [81,82]. The ratio of HI–QS/QI illustrated in Fig. 51 during the carbonization can be a measure of viscosity. The mesophase from LSVR showed a very low ratio in the final stage of the carbonizations, indicating that highly viscous mesophase is produced from the feedstock. In contrast, the HI–QS/QI of the mesophase from FCC-DO exhibited a very high value at the initial solidification stage, decreasing gradually with the progress of carbonization, indicating a moderate decrease in its viscosity. The mesophase from the mixture shows smaller but significant values of the ratio during solidification.

The carbonization temperature influences the value. The 7/3 mixture showed comparable ratios to those of FCC-DO alone. The 5/5 mixture showed similar levels of the ratio at 460°C with that of 7/3 mixture at 480°C. Thus, it can be seen that the mesophase from mixtures can be comparable in viscosity change to that from FCC-DO alone or much better than that from LSVR alone.

The chemical structure of HI certainly influences the properties of the mesophase. The accuracy of the analyses at the present level is not good enough for detailed discussion, although some characteristics are observable as already described.

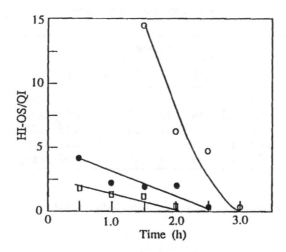

FIG. 51 Relation between carbonization time and ratio of HI–QS/QI of inter-mediates produced at 480°C under 8 kg/cm^2. ○, FCC-DO; ●, 7/3 mixture; □, LSVR.

In the second stage, appropriate gas evolution is required to rearrange the mesogen molecules uniaxially along the axis during the solidification process where the viscosity of the mesophase increases. The FCC-DO certainly gives methyl groups attached to the aromatic rings as a source for such a gas to give an excellent needle coke. The LSVR has many such gas sources since it carries a number of noncarbonized aromatics, alkyl side chains, and straight chain paraffins.

Table 15 summarizes the rate of gas evolution during the carbonization of the respective feedbacks under 8 kg/cm^2 at 480 and 460°C. The regions of solidification are also indicated. The rate of gas evolution from FCC-DO at 480°C increase sharply up to 30 min when the rate is maximum and then decreases sharply to zero at 3.5 h. The rate of gas evolution at 2.5–3.0 h, which is the region of solidification, decreases from 12 to 8 ml/min. The rate of gas evolution from LSVR at 480°C increases up to 20 min to reach the maximum rate and then decreases. The rate of gas evolution at solidification decreases from 26 to 10 ml/min. The rate of gas evolution from the 7/3 mixture at 480°C also increases up to 20 min to reach the maximum rate and then decreases.

TABLE 15 Rate of Gas Evolution (ml/10 min)

Temperature		480°C				460°C	
Time					Time		
(h)	(min)	FCC-DO	7/3	LSVR	(h)	7/3	5/5
	10	40	76	62	1.0	24	22
	20	88	102	116	2.0	36	56
0.5	30	98	78	70	3.0	32	42
	40	76	64	46	4.0	28	38
	50	54	58	42	5.0	20	32
1.0	60	46	48	42	6.0	18	22
	70	42	38	38	7.0	16	16
	80	36	30	32	8.0	16	12
1.5	90	28	26	26	9.0	12	10
	100	24	22	24	10.0	8	8
	110	20	20	22	10.5	4	6
2.0	120	16	18	10	11.0	4	6
	130	14	16	6	11.5	2	2
	140	14	16	0	12.0	2	0
2.5	150	12	12		12.5	0	
	160	12	8				
	170	12	0				
3.0	180	8					
	190	6					
	200	0					

▨ , Solidification zone (80).

The rate of gas evolution at solidification decreases from 18 to 12 ml/min.

Although the rates of gas evolution at the solidification point from both 7/3 and 5/5 mixtures in the carbonization at 460°C are much smaller than those at 480°C, the 5/5 mixture produces more gas during solidification. The sources of evolved gases are proposed to be paraffins and alkyl side chains of aromatics in the HS and HI–BS fractions, which give light hydrocarbons. Such sources are certainly plentiful in LSVR and stay in the mesophase before solidification. Hence, the mixture inherits such sources from the partner LSRV, and as expected and observed, more gas is evolved during solidification, which is favorable for the uniaxial arrangement.

It is of value to discuss why a particular mixing ratio of FCC-DO and LSVR is best at a particular carbonization temperature. The LSVR is more reactive at a higher temperature, requiring more FCC-DO to develop excellent mesophase. More gas evolution during solidification is expected at a higher carbonization temperature. The LSVR becomes much less reactive at a lower temperature, requiring lesser amounts of FCC-DO for moderation of its reactivity to produce a fine bulk mesophase; however, more sources for gas evolution are necessary for the uniaxial rearrangement, since the carbonization is slowed. Thus, the mixing ratios of 7/3 and 5/5 can be best at 480 and 460°C, respectively.

In summary, the cocarbonization of two feedstocks can control mesophase formation and provide more gas at its solidification, producing a better coke than those from each feedstocks alone. The mixing ratio and the carbonization temperature cooperatively influence the carbonization process through the reactivity, solubility, and hydrogen-donating ability of carbonizing molecules in both feedstocks. Thus, a particular carbonization temperature defines a particular mixture ratio for the best coke. The hydrogen-donating ability is often related to the naphthenic structure of the feed.

C. Future Application of Cocarbonization

Cocarbonization is not yet fully clarified in its effectiveness and mechanisms. Better additives, by careful selection and enhancement of their modifying activity, may facilitate greater efficiency and applicability of cocarbonization to needle coke production.

Cocarbonization of ethylene tar pitch (ETP) and coal tar pitch leading to better needle coke has now started to attract some interest to solve problems such as too high a reactivity of one component, and impurities of oxygen- and nitrogen-containing compounds in another component, which, respectively, cause an increase of CTE and puffing [82a].

XI. MOSAIC COKE OF LARGE CTE PRODUCED AT THE BOTTOM OF THE COKER DRUM

A. Problems of Bottom Coke

It has been recognized in the commercial operation of delayed coking that cokes of poor quality are often, but not always, produced at the bottom of the coker drum [83–87]. It should be pointed out that the quality of the needle coke produced as a major product and the amount of bottom (lower quality) mosaic coke are both taken into account when the carbonization process and feedstock are evaluated because both factors influence the total quality of cokes produced in a commercial coker drum.

Shot coke problems arising from the delayed coking of heavy residue to produce principally a cracked oil with the least amount of cheap fuel coke by-product have been known for some time. Shot cokes not only deteriorate the value of solid fuel, because they are difficult to grind, but are also hazardous during the drum decoking step, because of unexpected falling of hot shots from the drum.

Since these two relevant problems originate from similar causes, it is worthwhile examining the phenomena and mechanisms associated with the formation of bottom mosaic and shot coke, both of which can be studied using tube bombs.

B. Bottom Coke of Fine Mosaic Texture

As described previously, LSVR and its mixtures with FCC-DO produced areas of mosaic texture at the bottom of the coke lump. Figure 52 shows a montage of photographs of the whole area of a coke lump produced from a mixed feed (blend ratio: 7/3; FCC-DO/LSVR) at 480°C and 8 kg/cm^2 (A), and microphotographs of the middle and bottom parts in larger magnification (B–D). There are various types of flow texture that are uniaxially oriented (C) and not oriented (B) at the middle part of the lump, whereas the mosaic texture is located at the bottom part and is clearly distinguishable from the flow texture in the middle part. The thickness of the mosaic part at the bottom can be measured easily in the montage photograph.

The thickness of bottom mosaic belts are plotted against the mixing ratio in Figure 53, where the carbonization temperatures are 480 and 460°C, respectively. The thickness is about 1 mm in the cokes produced at 480°C from LSVR alone and its mixtures regardless of the mixing ratio. The thickness is much larger in the cokes produced at 460°C. The LSVR alone at the latter temperature gives a mosaic belt thickness of 3 mm, and the thickness decreases by adding more FCC-DO to the mixture. The mosaic belt thickness is zero in the coke from the FCC-DO alone.

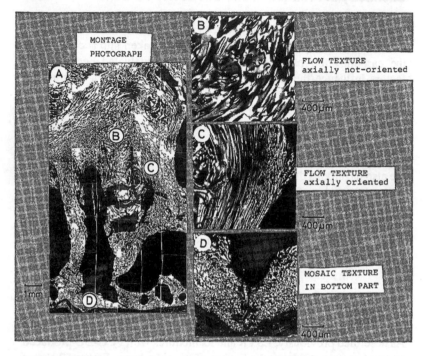

FIG. 52 Montage photographs and photomicrographs of the coke lump produced at 480°C under 8 kg/cm^2 from mixture of FCC-DO/LSVR (blending ratio = 7/3).

Carbonization of mixtures (FCC-DO/LSVR; 7/3 and 5/5) was studied at a higher pressure (16 kg/cm^2) and/or at a higher temperature (490°C) to reduce the amount of bottom mosaic coke. As illustrated in Fig. 54, the bottom mosaic coke becomes thinner (height 0.5 mm) when the carbonization temperature and pressure are higher, although it is still certainly observable. The uniaxial arrangement of the flow texture is slightly deteriorated by increasing the pressure when cokes produced at the same carbonization temperatures are compared. The higher temperature appears to increase the areas of turbulent flow texture as pointed out in the figure.

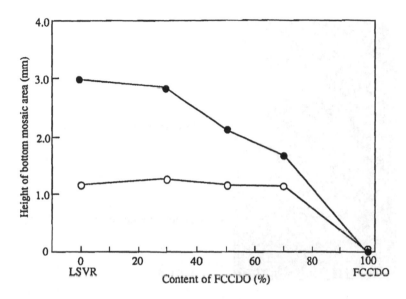

FIG. 53 Relation between the content of FCC-DO and height of bottom mosaic texture. Carbonization temperature: ○, 480°C; ●, 460°C. Carbonization pressure 8 kg/cm².

C. Mechanism of Bottom Mosaic Formation

Bottom mosaic coke is produced at the very earliest stages of the carbonization. The most reactive fractions of the asphaltenes in LSVR may be condensed in the early stages of carbonization, losing alkyl side chains and being converted into poly-condensed aromatic molecules in the micelle. Such polyaromatic hydrocarbons are hardly soluble in the paraffin-rich matrix and settle to the bottom of the vessel, where carbonization progresses in a highly viscous environment with little chance of mesophase growth and evolution of volatiles, giving fine mosaic cokes of high density.

Such a mechanism may also operate in cocarbonization. Although a major part of the most reactive asphaltene may be dissolved and moderated by fractions of the FCC-DO in the feed, some of the most reactive portions still follow the above mechanism to give bottom mosaic coke, but in a lesser amount. The amount of such a highly reactive

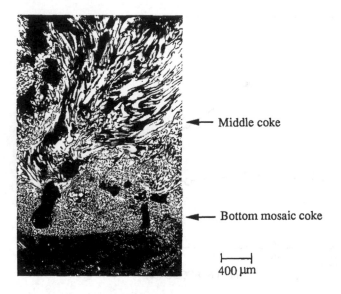

← Middle coke

← Bottom mosaic coke

├────┤
400 μm

FIG. 54   Micrographs of bottom mosaic coke produced at 490°C in tube bomb from a blend of LSVR/FCC-DO=50/50.

portion may strongly depend on the carbonization conditions and mixing ratio. A higher temperature (up to 490°C) increases solubility of this portion in FCC-DO and enhances the hydrogen-donating ability of FCC-DO, resulting in a decrease in the amount of bottom mosaic coke. Such conditions are not necessarily best for the uniaxial arrangement of flow texture in the carbonization of the major portion as discussed in a previous section. Furthermore, too high a temperature may accelerate the reactivity of the asphaltene to increase the amount of bottom coke. At lower temperatures, the cocarbonization of LSVR may operate with less of the FCC-DO because of limited solubility and hydrogen-transferring ability, although the reactivity of the asphaltene may be reduced, requiring some FCC-DO to dissolve or moderate the source of bottom coke.

According to the mechanisms described previously, better FCC-DO can be selected as a more favorable partner for LSVR. The LSVR can be also modified as a more acceptable component of the cocarbonization. Better additives may also be found in the refinery and coking industries. In the tube bomb study, the influences of recycle feed are neglected. Its favorable contribution to commercial coke production appears to be significant.

D. Shot Coke

Shot cokes are produced from some feedstocks in a commercial delayed coker. This spherical coke of high density shows a very fine mosaic structure at its core and consists of slightly larger mosaic units at its thick skin, which are roughly oriented in an onionlike arrangement. The coke particles often agglomerate, appearing like a bunch of grapes.

The structure of shot coke suggests a mechanism for its formation. A highly reactive portion in the feedstock first forms poly-condensed aromatics that become insoluble and separate from the matrix. These act as cores around which the shot coke carbonaceous spheres grow. The carbonization proceeds to give a very fine mosaic texture with a high density because little volatile matter is released. The spherical coke particles may move around in the matrix because of turbulent flow in the coke drum, collecting the condensing aromatic molecules on their surface. The rapid rotation of the spheres may allow accumulation of layers in an onion-skin alignment. The rapid turbulence and low viscosity may enlarge the size of other spheres. Grown spheres may fall to and deposit on the wall of the bottom of the coker vessel where the turbulent flow may not reach. Thus, shot coke is mostly found in the bottom of the coker drum. However, when the viscosity of the coking substance is high enough to suspend the core spheres until completion of the coking process, shot cokes may be found anywhere in the drum. The former case is possible when a limited amount of very heavy fraction is present in a rather paraffinic matrix and the latter when the feed consists of high boiling fractions of poor mutual solubility.

According to the mechanism described, phase separation of the shot coke core is the first stage in its formation. Hence, dissolution of this core may prohibit or at least decrease its formation. Cracked oil of high aromaticity may dissolve the core immediately after its formation. Hydrogen transfer may be very effective to prohibit. Again, there is more chance of its formation at the bottom of the drum where no additional cracked oil is expected to be found. Very limited amounts of aromatic hydrocarbons in the cracked products, or low solubility of the core fraction formed from the heaviest components of the feed, may not allow any way to reduce the formation of shot coke.

At present, detailed relationships between the formation of shot coke and the components of the matrix are not available. Further study is necessary. Nevertheless, it is suggested that phase separation of the heaviest fraction should be avoided to suppress the formation of shot coke. Hence, the elimination of the heaviest fraction from the coking feed and enrichment of it with solvent aromatic fractions, as well as hydrogen donor reagents, may be keys to solving the problems.

XII. SELECTION, ADJUSTMENT, AND MODIFICATION OF FEEDSTOCKS FOR BETTER COKES

A. Better Feed for Needle Coke

Although the optimum conditions may give the best coke from any feed, starting with an optimized feed should lead to the best needle coke. There are several approaches to optimizing the feed for carbonization as well as cocarbonization.

1. Selection of better feed and additive. There are many varieties of residual oils for delayed coking in the refinery and the coking industries, such as straight vacuum residua of crudes; pyrolyzed, hydrocracked, and cat-cracked tars; extracted residues; coal tars; and liquefied coal products. Furthermore, each of these vary according to the origins and conditions of its production. Table 16 summarizes some properties of FCC-DO produced in Japanese refineries [88a]. The extent of their variabilities are clear; hence, the selection of feed or additive is the first priority.
2. Adjustment. A fraction in the feed can be a poison for coking or remain in the coke as an unwanted contaminant. Thus, the removal of these types of fraction may provide a chance to obtain better feeds. Addition of suitable additives may be another approach based on the cocarbonization results discussed earlier.
3. Modification. Structural modification of a part or whole of any feed can be, in principle, convert it to an optimum feed for needle coke production.

Thermal and catalytic processes for cracking, aromatization, hydrogenation, hydropurification, and hydrocracking are applicable. Catalytic reactions are the most selective and powerful for these types of conversion; however, the residual oil may contain catalyst poisons, and high pressures are often necessary in the reactions. Thus, the catalyst

TABLE 16 Properties of FCC-DO and LSVR Produced in Japanese Refineries [88a]

	A1	A2	B	C	D	E	F	G	LSVR
sp gr[a]	1.038	1.005	1.065	1.0372	0.9881	0.9621	0.9954	0.9930	0.928
Viscosity (cSt)	102.8	45.35	12.98	25.4	25.1	24.6	16.1	13	74.8
Elemental analysis (%)									
C	89.60	89.52	90.63	89.76	88.90	89.19	89.86	89.64	86.2
H	9.05	9.81	7.80	8.37	9.93	10.43	9.38	9.50	12.5
N	0.06	0.08	0.10	0.22	0.19	0.07	0.14	0.04	0.4
S	0.87	0.34	1.08	1.38	0.43	0.12	0.35	0.35	0.2
O	0.42	0.25	0.39	0.27	0.55	0.19	0.27	0.47	0.7
Conradson C (%)	5.54	4.45		4.12	4.03	2.50	1.66	2.83	10.2
f_a[b]	0.62	0.58	0.69	0.67	0.57	0.55	0.61	0.59	0.18
σ_{al}[c]	0.36	0.34	0.36	0.35	0.36	0.33	0.35	0.36	0.52
C_{side}[d]	14.80	37.25	9.39	6.77	24.20	78.77	13.78	33.06	
Composition (%)									
saturate	23	30	4	8	27	39	31	36	52
aromatic									
light	52	55	44	75	50	27	43	10	27
heavy	24	14	51	16	22	33	25	52	11
resin	1	1	1	1	1	0	1	2	11
asphaltene	0	0	0	0	0	0	0	0	10

[a]Specific gravity.
[b]Carbon aromaticity.
[c]Number of alkyl side chains on aromatic rings in structural unit.
[d]Number of carbon atoms in alkyl side chains in structural unit.

and the catalytic process should be carefully designed to be economically viable.

B. Some Examples of Feed Selection

1. Coal Tar

Coal tar is a by-product of coal coking in an oven to produce blast furnace coke. The coal blend and coking conditions are varied to meet the supply of coal and demand of coke, thus changing the properties of the coal tar. Lower rank coal and lower coking temperature tend to give coal tar, which is a low-quality feed for needle coke as shown in Table 17 [88]. A relatively large number of alkyl groups on the aromatic rings, high oxygen functionality, and higher molecular weight of the tars are believed to be the causes of its higher reactivity and lower quality as shown in Table 18 [88]. Too high a coking temperature in the coking oven produces a tar that is chemically too stable and shows low gas evolution. Hence, the careful selection and storage of the tar is essential for the stable production of needle coke. Blending technology should be established when coal tar properties are so variable.

2. FCC-DO

The FCC-DO is a by-product of the FCC process, which produces a fairly aromatic oil. Nevertheless, the detailed structures of decant oil varies with the feed for the FCC process and the reaction severity is usually adjusted to achieve a given product. When gasoline production is maximized, the DO increases in aromaticity, which means less paraffins

TABLE 17 Properties of Cokes from Coal Tar Pitches

	A	B	C	D	E	F	G
CTE^a ($\times 10^{-6}/°C$)	0.8	1.3	0.4	0.7	0.2	0.7	0.4
RD^a (g/cm^3)	—	1.89	1.83	1.86	—	1.98	1.88
$N (wt\%)^a$	1.16	1.10	1.15	1.09	1.07	1.05	1.05
Coke yield ($wt\%)^b$	74	71	73	69	74	78	65

[a]Calcined cokes, at 1000°C, for 1 h.
[b]Carbonization condition; 500°C, 8 kg/cm^2.

TABLE 18 Properties of Coal Tar Feedstocks

Sample	A	B	C	D	E	F	G
Density	1.175	1.177	1.177	1.179	1.180	1.183	1.184
HS[a]	46	47	39	27	48	53	49
HI–BS	47	47	54	64	45	42	45
BI	7	6	7	9	7	5	6
C (wt%)	91.61	90.27	91.56	90.75	91.80	91.13	91.73
H (wt%)	5.19	5.12	5.17	4.98	4.97	4.92	4.95
N (wt%)	1.21	1.22	1.20	1.26	1.22	1.21	1.20
O (wt%)[b]	1.54	2.03	1.57	1.75	1.53	1.18	1.47
S (wt%)	0.43	0.50	0.46	0.48	0.47	0.50	0.43
C/H	1.47	1.47	1.48	1.52	1.54	1.54	1.54
O/C	1.26	1.69	1.29	1.45	1.25	0.97	1.54
f_{ac}	0.936	0.932	0.947	0.942	0.933	0.966	0.947
R_{nus}[c]	0.000	0.061	0.000	0.145	0.557	0.000	0.251
σ_{al}[c]	0.084	0.094	0.074	0.079	0.092	0.051	0.076
M_w[d]	302	313	323	333	366	370	324

[a]Solubility (wt%): HS; hexane soluble; HI–BS, hexane insoluble but benzene soluble; BI, benzene insoluble.
[b]Oxygen content was measured directly.
[c]Brown–Ladner analyses based on ^{1}H-NMR: f_a, carbon aromaticity; R_{nus}, number of naphthenic rings in the structural unit; σ_{al}, degree of substitution on aromatic ring in the structural unit.
[d]Analysis based on FS–MS.

and alkyl substituents. The sulfur content is principally derived from the feed. Hydrotreated feed for FCC often provides excellent DO feed for coking.

3. Ethylene Tar

Ethylene tar is highly aromatic and essentially free of heteroatoms. Nevertheless, the tar as received is not a suitable feed for coking. Its application is not recommended unless suitable adjustment or modification is made.

4. Straight Vacuum Residua from Crudes

The sulfur content in VR appears to be the most important concern for selection to obtain a low sulfur-containing coke. More detailed characterization may be necessary.

C. Adjustment

1. Coal Tar

It is well known that QI in coal tar pitch inhibits needle coke production. Its separation is performed by antisolvent precipitation. The oxygen and nitrogen compounds in the pitch are suspected to lower the coking properties and enhance the puffing problems, respectively. Hence, their removal is a target for study [89]. Selective extraction, using a series of particular solvents, and adsorption were investigated. The recovery of adsorbed species and regeneration of the absorbent are critical for the practical application of the latter approach.

2. VR

Asphaltene in the VR often is suspected of interfering with the coking and of increasing the impurity content in the resultant coke. Its removal may be effective [90]. It should be noted that the asphaltene never consists of single species. Selective removal of fractions that poison the reaction is required; otherwise, the coke yield will be very low.

3. FCC-DO

Too many paraffins in the FCC-DO, due to low severity of the FCC-process, cause the formation of bottom coke in cocarbonization [91]. Their removal may be effective with some DO, when its aromatic fraction does not carry too many long chain alkyl substituents [92]. Dealkylation of these long chain aromatics may be effective [93].

4. Ethylene Tar

Ethylene tar is not an excellent feed for coking. Some very reactive fractions in the tar are suspected to govern the carbonization. Hence, their removal may be effective [94].

D. Modification

1. Hydrogenation

Partially hydrogenated aromatic hydrocarbons are believed to be excellent hydrogen-donating agents that mediate the carbonization reaction in the mesophase development and gas evolution for the uniaxial rearrangement. Favorable results of partial hydrogenation are widely recognized for both the additive as well as the feed [95–97].

Hydrogenation-transferring modifications can be applicable. However, hydrogenation of aromatic rings in the coking feedstock of heavy hydrocarbons is not an easy task [98].

2. Hydrorefining

Removal of sulfur, nitrogen, and oxygen atoms from the feedstock is very useful for producing needle coke of a high quality in terms of anisotropic development and puffing inhibition [99]. However, it should be noted that the removal of such contaminants often leads to the degradation of aromatic rings, resulting, in the extreme, in no coke formation from the molecules after the removal of heteroatoms. In such cases, the content of heteroatoms in the feed is certainly reduced; however, less reduction of their content in the coke is achieved. It is necessary to ensure the condensation reactivity of the purified substrates.

Heteroatoms in the feed are believed to disturb the anisotropic development because they act as reactive sites for condensation and dehydrogenation agents [100]. It is very often true, nevertheless, that such reactivities are not due to their presence *per se* but to the detailed structures in which they are contained. It has been reported that some heterocyclic compounds produce excellent flow texture in the presence of aluminum chloride [101].

Since puffing can be solved by ways other than the elimination of heteroatoms, deep hydroclimination of such heteroatoms is not the only solution (Section V).

3. Hydrocracking

Long chain paraffins, too many alkyls substituents, and polymeric asphaltene disturb the carbonization. Such species are hydrocracked into harmless forms.

4. Acidic Cracking and Condensation

Cracking of paraffins and dealkylation from alkyl aromatics can be performed by acidic catalysts such as zeolites [102]. Primary condensation of smaller aromatic hydrocarbons is useful in controlling the carbonization of the feed. Aluminum chloride [103] and HF/BF$_3$ [104] are known to promote condensation at moderate temperatures without extensive dehydrogenation, producing naphthenic structures. The product can be an excellent additive.

The recovery and reuse of such catalysts are essential for practical applications. In terms of catalytic activity at lower temperatures, HF/BF$_3$

is much better than aluminum chloride and can be easily and completely recovered for reuse [104].

Solid catalysts such as zeolites are desirable for an industrial process; however, their activity appears to be too low at a lower temperature where no coking and dehydrogenation take place [105].

5. Thermal Process

Carbonization consists of a series of complex reactions that may proceed successively and consecutively according to the reactivity of the feed and reaction conditions (heating rate, temperature, and pressure for carbonization). Hence, selective reaction under the influence of particular cations can convert some feeds into better ones [106]. Pyrolytic cracking and condensation are major reactions, which are designed for the particular species concerned in the feed stock. The reaction conditions should be strictly controlled.

The pyrolysis of vacuum gas–oil is known to produce thermal tar as an excellent feed for needle coke as described in a previous section [107]. A highly aromatic feed of low sulfur content is produced.

It has been proposed that the thermal process can be used to adjust feed composition. For example, the most reactive portions are converted by the thermal process into a low quality coke that is removed in the following separation process [108], leaving the other components "purified" and better suited to make needle coke.

E. Problems of Hydrotreating Process of Feed Modification

Hydrotreating processes are all very useful in the refining of residues. However there are several problems to be considered. First of all, the high hydrogen pressure of the process raises the cost of the process considerably, especially when the size of the facility is limited due to the restricted amount of coke production. A catalyst that is active at lower than conventional pressures and that is resistant to poisons in the residual oil is strongly desired. Hydrogen consumption, which raises the operating cost, should be saved as much as possible.

When the process is designed using currently available Co–Mo or Ni–Mo catalysts, catalyst deactivation by coke deposition on the catalyst, which lowers the catalyst of life, is of most concern. The catalysts should be tolerant to coke deposition and should not cause coke formation. Pore structure, acidity, and polarity of the catalyst supports are carefully designed.

Design of a series of catalytic steps is very useful. One of the present authors has proposed a two-stage hydrotreating process: extensive hydrogenation in the first step at a relatively low temperature followed by extensive hydrocracking or hydrorefining at relatively high temperatures. The border between low and high temperature appears to be at 400–410°C where the hydrogenation/dehydrogenation equilibrium may exist for the aromatic rings [109]. Most suitable catalysts can be chosen to meet the requirements of the respective steps [110].

It is also important to select the solvent for the process. Some solvents are very effective for suppressing coke deposition [111] on catalyzed surfaces.

A combination of all such ideas may moderate the reaction conditions and accelerate the reaction, although more controlled systems are necessary. A lower cost of upgrading facilities can be also expected in multi-streams of refining for plural objectives.

XIII. CALCINATION OF GREEN COKE INTO COMMERCIAL NEEDLE COKE

A. What Happens in the Calcination Stage?

Calcination is the final step in the production of commercial needle coke. Preliminary graphitization and purification into pure carbon are expected to occur at this stage.

The key step in calcination has been considered rather simply to be the adjustment of the amount of volatile matter by control of the final calcination temperature for commercial coke. However, a recent study revealed that this step is very critical for increasing the quality of the coke in terms of its crack development due to extensive shrinkage, content of heteroatoms, and surface chemistry. Such structural factors are very influential on the CTE, puffing trend, strength (grain size), and wettability of the coke [112,113].

Thus, the sensitive control of heating rate, final temperature, calcination atmosphere in the temperature range, and movement in the kiln during the calcination are of great interest.

Carbonaceous substances, including green coke, release most of their hydrogen in the temperature range of 700–800°C, starting with the preliminary ordering as illustrated in Fig. 13, although a detailed structure is not yet clear because the carbon at this stage is still noncrystalline [114].

Properties of carbon such as electrical conductivity, CTE, and density change vary drastically in this temperature range as illustrated in Figs. 21, 55, and 56 and gradually above this temperature range, eventually reaching those of graphite [115,116].

B. Crack Development During Calcination and Its Influence on Coke Properties

The drastic change in density during calcination causes a large shrinkage [117–119]. The densification due to preliminary graphitization takes place in both directions along the *a*- and *c*-axes. However, the extents are anisotropic, with that along the *c*-axis being much larger. The shrinkage, thus induced, takes place within the respective grains of the same layer orientation (reflected by the optical anisotropy) as expected from the restricted atomic movement to form the graphite layers. Hence, cracks develop between the grains. The model structure of needle coke illustrated in Fig. 16 illustrates this manner of crack development.

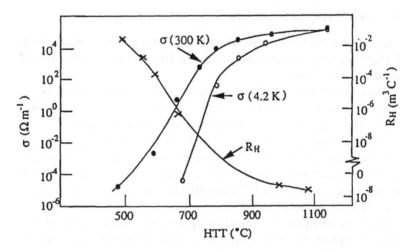

FIG. 55 Variation of conductivity and Hall coefficient for an anthracene char as a function of heat-treatment temperature. The Hall coefficient changes from positive for HTT ≤700°C to negative for HTT ≤900°C. Figure adapted from Ref. 4.

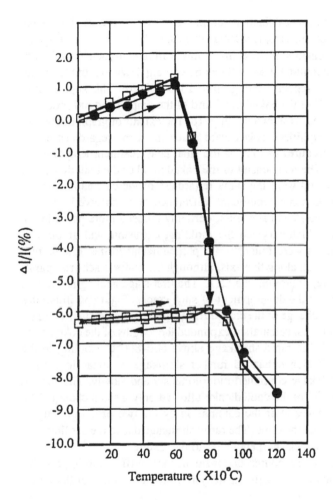

FIG. 56 Dimensional changes of coke during calcination. ●, traditional method; ▫, new method.

The extent of crack development is controlled by the atomic movement that is aided by the gas evolution as defined by the temperature and directed by the crystallization. Thus, the temperature of gas-release during the calcination step is very influential on the shrinkage. Since the temperature for the calcination is usually increased to 1400°C, the heating program in this temperature range is critically important.

Cooling from the calcination temperature causes cracks due to the stress induced by anisotropic thermal shrinkage. Koa Oil has investigated a two-stage calcination process involving heating up to 800°C and holding at this temperature for some time and cooling down to room temperature followed by regular heating up to 1300°C [120]. This significantly reduces the CTE of the baked mold and of the graphitized rod produced from needle cokes, as shown in Table 18. The extents of graphitization of calcined and graphitized cokes were much the same regardless of the calcination procedures as long as the final heat-treatment temperatures were the same. The real density of the baked mold or the bulk density of the graphitized rod were the same regardless of the calcination procedures; however, some significant difference was observed in their porosity. The two-stage calcination certainly induced more pores in the calcined coke, which survived the molding, extrusion, and graphitization. Some cracks were found to run perpendicular to the flow texture due to densification along this axis, although the major cracks run parallel to the texture, improving the CTE. The first-stage heating is claimed to be essential for the development of such cracks. Figure 56 illustrates the dimensional change of the coke during regular and two-stage calcination. In both, the first calcination expanded the lengths of the cokes up to 600°C, which led to rapid shrinkage above 600–800°C. Cooling in the two-stage calcination enhanced further shrinkage, whereas in regular calcination shrinkage continued monotonously and slowly. The second heating to 1300°C of the cooled coke allowed only a small change. The lengths of the cokes after the calcination by the two procedures up to 1000°C were much the same. The rapid shrinkage due to the cooling may induce the microcracks perpendicular to the texture.

The calcination procedures also influence the puffing during graphitization. The microcracks (1 μm in size) may provide a route for liberating heteroatoms.

The calcination procedure also influences the mechanical strength of the coke. Table 19 indicates the increased strengths of rods prepared from coke calcined through two stages. Shell claimed that a similar calcination increased the grain size of the resultant coke [121]. A similar mechanism may work best here where the large cracks leading to breaking of the coke grains are suppressed by the development of many microfissures. On the other hand, rapid heating in the rotary hearth is suspected as a reason for severe development of large cracks.

TABLE 19 Properties of Calcined Needle Coke Produced by Conventional and Two-Stage Procedure

	A		B	
Coke calcination procedure	Two	Conventional	Two	Conventional
Baked rod[a]				
CTE^c ($\times 10^{-6}$°C)	1.5	1.9	1.2	1.6
Real density	2.094	2.088	2.095	2.092
Porosity (%)	40.3	35.8	37.0	33.1
d_{002} (Å)	3.444	3.440	3.447	3.446
Lc (Å)	38	40	38	40
Graphitized rod[b]				
CTE^c ($\times 10^{-6}$°C)	1.0	1.2	0.8	1.0
Bulk density	1.52	1.53	1.52	1.52
Bending strength (kg/cm^2)	115	110	115	105
Young's modulus (kg/cm^2)	790	740	810	740
d_{002} (Å)	3.366	3.366	3.365	3.364
Lc (Å)	840	880	1000	1000

[a]Baked at 1000°C.
[b]Graphitized at 2800°C.
[c]CTE measurement 30–100°C.

C. Removal of Heteroatoms During Calcination

Cokes liberate their heteroatoms through a thermal process during calcination, with more being liberated at higher temperatures. Hence, calcination at 1400–1500°C removes most nitrogen remaining in the coke, preventing the puffing problem [122]. However, such calcination removes many surface functional groups, drastically reducing the wettability. Thus, calcination at higher temperatures has the disadvantage of not only larger energy consumption but also lower surface energy of the coke.

Desulfurization at the calcination stage is enhanced by hydrogen molecules. The contact of hydrogen with sulfur in the coke is the key step for this reaction, microcracks defining the effective surface area and major structural factors. Hence, such desulfurization is most rapid in the temperature range of 700–800°C [123].

Oxidation may take place during the calcination when any oxygen leaks into the high-temperature zone in the kiln or hearth. Diffusion-controlled gasification takes place, thus increasing the large pores in the coke and reducing the strength of the coke. However, controlled gasification at lower temperature may induce micropores for better CTE and less puffing.

XIV. CONCLUSIONS AND FUTURE TASKS TO IMPROVE NEEDLE COKE PRODUCTION

Massive amounts of premium needle coke, a very essential feed for electrodes and other carbon artifacts, have been produced for 40 years in delayed cokers. Its quality and cost have improved gradually, being based principally on empirical trial-and-error approaches. Hence, the technology stays at the state of the art. More logical approaches to ideal coke and its production are needed now and in the next decade, when a huge amount of scrap should be more economically and efficiently recycled.

In the present review, the authors intended to review several basic aspects of the chemistry relevant to needle coke production. Carbonization in the tube bomb simulates coking in the commercial coker well, allowing mechanistic studies in the laboratory. The precise recognition of differences in two coking procedures may provide clues and key information to gain further precise understanding of coking in delayed coker drums where complex physical, chemical, and transport phenomena take place simultaneously.

More extensive trials and studies of cocarbonization using blends of more kinds of feeds appear promising. Extensive modification of feeds of a wide variety of origins may be possible when more efficient processes and more powerful catalysts are developed for heavy hydrocarbons. Thus, the economy and technology of coking should match those of the whole hydrocarbon industry, especially the fuel and energy industry. In turn, coking technology can contribute novel ideas to the industry.

Simpler decoking is another major target for extensive development. The materials of the walls of the drum may be critical for easy decoking. Substitutes for the coker drum can also be investigated. Injection of mesophase pitch into the tips of needle coke may be possible.

ACKNOWLEDGMENTS

The authors (N. O. and K. F.) are grateful to Koa Oil and Nippon Steel Chemical Co., Ltd., respectively for permission to publish this review.

REFERENCES

1. M. C. Sooter, *IX Polish Graphite Conference*, 1988, p. 137.
2. International Iron and Steel Institute, *The Electric Arc Furnace*, Section 7, Committee on Technology, Brussels, 1981.
3. L. L. Winter and E. L. Piper, *Encyclopedia of Chemical Technology*, Vol. 4, (Kirk and Othmer, eds.), 1982, pp. 570–576.
4. M. B. Redmount, *Proceedings of Met. Bull. 2nd International Mini Mill Congress*, 1984, p. 158.
5. C. R. Taylor, *Electric Furnace Steelmaking*, Iron and Steel Society, 1985, p. 71.
6. H. Matsumura, Nishiyama Memorial Award Lecture, Japan Iron and Steel Institute 109 (1986).
7. H. A. Taylor Jr., *Ceram. Bull, 67*(5), 862 (1988).
8. V. P. Sosedov, V. V. Mochalov, and A. K. Sannikov, *Proceedings of 9th Polish Graphite Conference*, 1988, p. 1.
9. Y. Yin, T. J. Mays, and B. McEnaney, *Carbon 27*, 113 (1989).
10. R. DeBiase, J. D. Elliott, and T. E. Hartnett, *Petroleum-Derived Carbons* ACS Symposium Series No. 303 (1986), p. 155.
11. H. H. Brandt, *Petroleum-Derived Carbons*, ACS Symposium Series No. 303 (1986), p. 172.
12. H. Murata, T. Miyazaki, H. Hiroaka, J. Koide, S. Hirage, Y. Tashiro, W. Migitaka, and M. Iwasa, *Ext. Abstracts Int. Symp. on Carbon*, Toyohashi, 1982, p. 5.
13. R. E. Franklin, *Proc. Roy. Soc. A 209*, 96 (1951).
14. H. Marsh and J. A. Griffiths, *Ext. Abstracts Int. Symposium on Carbon*, Toyohashi 1982, p. 81.
15. I. Michida and Y. Korai, *Nenryo Kyokaishi 64*, 796 (1985).
16. I. Mochida, T. Oyama, Y. Nesumi, and Y. Korai, *Carbon 25*, 259 (1987).
17. I. Mochida, K. Maeda, Y. Korai and K. Takeshita, *Nenryo Kyokaishi 61*, 1050 (1982).
18. I. Mochida, M. Ogawa, and K. Takeshita, *Bull. Chem. Soc. Japan 49*, 514 (1976).
19. B. T. Kelly, *Physics of Graphite*, Applied Science Publishers, 1981, p. 19.
20. J. W. Patrick and A. E. Stacey, *Fuel 54*, 256 (1975).
21. J. E. Zimmer and J. L. White, *Ext. Abstracts 13th Bien. Conf. on Carbon*, Irvine, CA, 1977, p. 318.

22. R. Yamashita, H. Nagino, K. Fujimoto, and K. Ohtsuka, *High Temp.-High Press.* 455 (1988).
23. F. M. Collins, *Proceedings of 1st and 2nd Conferences on Carbon*, Pergamon, New York, 1956, p. 177.
24. M. P. Whittaker and L. I. Grindstaff, *Carbon 7*, 615 (1969).
25. E. Fitzer and S. Weisenburger, *Rev. Int. High Temp. & Refract. 12*, 69 (1975).
26. I. Letizia, *High Temp.-High Press. 9*, 297 (1977).
27. M. H. Wagner, W. Hammer, and G. Wilhelmi, *High Temp.-High Press. 13*, 153 (1981).
28. I. Letizia, A. D. Pasquale, and F. Calderone, *High Temp.-High-Pres. 13*, 303 (1981).
29. R. Yamashita, T. Nishihata, K. Kawashima, and K. Fujimoto, *Proceedings of Japan Carbon Society Annual Conference*, 1988, p. 132.
30. E. Fitzer, D. Kompalik, and O. Wormer, *Proceedings of Carbon '86*, Baden-Baden, Germany, 1986, p. 116.
31. K. Fujimoto, M. Yamada, M. Iwasa, and H. Nagino, *High-Temp.-High Press. 16*, 669 (1984).
32. K. Fujimoto, M. Sato, M. Yamada, R. Yamashita, and K. Shibata, *Carbon 24*, 397 (1986).
33. I. Mochida and H. Marsh, *Tanso 126*, 86 (1986).
34. K. Fujimoto, K. Ohstuka, M. Iwasa, and R. Yamashita, *High Temp.-High Press. 16*, 677 (1984).
35. M. H. Wagner, H. Jager, I. Letizia, and G. Wilhelmi, *Proceedings of 18th Biennial Conference on Carbon*, Worcester, MA, 1987, p. 38.
36. E. A. Heintz, *Ext. Abstracts 5th Int. Conf. Carbon and Graphite*, London, (1978) p. 575.
37. S. R. Brandtzaeg and H. A. Oye, *Carbon, 26*, 163 (1988).
38. I. Mochida, T. Oyama, and K. Fujimoto, *Tanso 131*, 187 (1987).
39. R. T. Lewis, *Ext. Abstracts 18th Biennial Conference on Carbon*, Worcester, MA, (1987), p. 36.
40. E. Fitzer and H-P. Janoschek, *High Temp.-High Press. 10*, 527 (1978).
41. H. Preiss, H. Mehner and P-M. Wilde, *High Temp.-High Press. 16*, 309 (1984).
42. E. G. Morris, K. W. Tucker, and J. A. Joo, *Ext. Abstracts 16th Biennial Conference on Carbon*, San Diego, 1983, p. 595.

43. K. Fujimoto, M. Yamada, R. Yamashita, and H. Nagino, *High Temp.-High Press.* *19*, 687 (1987).
44. I. Mochida, Y. Korai, Y. Nesumi, and T. Oyama, *Ind. Eng. Chem. Prod. Res. Dev.* *25*, 198 (1986).
45. I. Mochida, T. Oyama, Y. Nesumi, and Y. Korai, *Carbon 25*, 259 (1987).
46. J. D. Brooks and G. H. Taylor, *Carbon 3*, 185 (1965).
47. J. D. Brooks and G. H. Taylor, in *Chemistry and Physics of Carbon*, Vol. 4 (P. L. Walker, Jr. and P. A. Thrower, eds.), Dekker, New York, 1968, p. 254.
48. I. Mochida, T. Oyama, and Y. Korai, *Carbon 26*, 49 (1988).
49. H. Marsh and P. L. Walker, Jr., in *Chemistry and Physics of Carbon*, Vol. 15 (P. L. Walker, Jr., and P. A. Thrower, eds.), Dekker, New York, 1979, p. 228.
50. J. L. White, *Petroleum Derived Carbons*, ACS Symposium Series, No. 21, (1976) p. 282.
51. M. Kakuta, T. Tsuchiya, H. Tanaka, and K. Noguchi, *Ext. Abstracts 16th Biennial Conf. Carbon*, San Diego, 1983, p. 16.
52. I. C. Lewis, *Carbon 18*, 191 (1980).
53. D. O. Rester and C. Rowe, *Carbon 12*, 18 (1974).
54. I. Mochida, Y. Q. Fei, T. Oyama, and Y. Korai, *J. Mater. Sci. 22*, 3989 (1987).
55. I. Mochida, T. Oyama, Y. Q. Fei, T. Furuno, and Y. Korai, *J. Mater. Sci. 23*, 298 (1988).
56. I. Mochida, T. Oyama, Y. Korai, and Y. Q. Fei, *Oil & Gas J.* May 2, 73 (1988).
57. I. Mochida, T. Oyama, Y. Korai, and Y. Q. Fei, *Fuel 67*, 1171 (1988).
58. J. K. Brown and W. R. Ladner, *Fuel 39*, 79 (1960).
59. Heavy Oil Division, Refining Section of the Japan Petroleum Institute, *J. Japan Petrol Inst. 24*, 44 (1981).
60. K. S. Seshadri, E. W. Albaugh and J. D. Bacha, *Fuel 61*, 336 (1982).
61. D. W. Van Krevelen and H. A. G. Chermin, *Fuel 33*, 79 (1954).
62. T. F. Yen, J. G. Erdman, and S. S. Pollack, *Anal. Chem. 33*, 1587 (1960).
63. I. Mochida, A. Azuma, and Y. Korai, *Ext. Abstracts 19th Bien. Conf. on Carbon*, Penn State Univ., PA, 1989, p. 184.
64. I. Mochida, K. Maeda, and K. Takeshita, *Carbon 15*, 17 (1977).
65. J. L. White and M. Buechler, Preprints, Division of Petroleum Chemistry, *Amer. Chem. Soc. 29*, 1984, p. 387.

66. I. Mochida and Y. Korai, Abstracts, Int. Symp. on Science and New Applications of Carbon Fibers, Toyohashi, Japan (1984).
67. R. J. Diefendorf, *Ext. Abstracts 16th Bien. Conf. on Carbon*, San Diego, Plenary Lecture, (1983), p. 28.
68. I. Mochida, T. Oyama, Y. Korai, and Y. Q. Fei, *Oil & Gas J.* 73 (1988).
69. I. Mochida, T. Oyama, Y. Korai, and Y. Q. Fei, *Fuel 67*, 1171 (1988).
70. H. Marsh and P. L. Walker, Jr., in *Chemistry and Physics of Carbon*, Vol. 15 (Philip A. Walker, Jr. and Peter A. Thrower, eds.), Dekker, New York, 1979, p. 228.
71. H. Marsh, I. Macefield, and J. Smith, *Ext. Abstracts 13th Bien. Conf. on Carbon*, Irvine, CA, 1977, p. 21.
72. I. Mochida, K. Amamoto, K. Maeda, and K. Takeshita, *Fuel 56*, 49 (1977).
73. I. Mochida, K. Amamoto, K. Maeda, and K. Takeshita, *Fuel 57*, 225 (1978).
74. Y. Korai and I. Mochida, *Fuel 62*, 893 (1983).
75. I. Mochida, H. Toshima, Y. Korai, and T. Matsumoto, *J. Mat. Sci.* 24, 2191 (1989).
76. I. Mochida, H. Toshima, Y. Korai, and T. Matsumoto, *J. Mat. Sci.* 23, 670 (1988).
77. I. Mochida, H. Toshima, Y. Korai, and T. Naito, *J. Mat. Sci.*, in press.
78. C. A. Stokes and V. J. Guercio, *Erdöl & Kohle Erdgas Petrochemie* 38, 31 (1985).
79. I. Mochida, Y. Korai, T. Oyama, Y. Nesumi, and Y. Todo, *Carbon* 27, 359 (1989).
80. Y. Nesumi, Y. Todo, T. Oyama, I. Mochida, and Y. Korai, *Carbon* 27, 367 (1989).
81. J. A. Roetling, J. J. Gebhardt, and F. G. Rouse, *Carbon 25*, 23 (1987).
82. M. Kakuta, N. Tsuchiya, and M. Kohriki, *Tanso 85*, 55 (1976).
82a. I. Mochida, Y. Q. Fei, Y. Korai, and T. Oishi, *Fuel 69*, 672 (1990).
83. M. Kakuta, T. Tsuchiya, and M. Kooriki, *Tanso 85*, 55 (1976).
84. S. Ester, R. G. Jenkins, and F. J. Derbyshire, *Carbon 24*, 77 (1986).
85. I. Mochida, T. Furuno, Y. Korai, and H. Fujitsu, *Oil & Gas J. 3*, 51 (1986).
86. I. Mochida, T. Oyama, Y. Q. Fei, T. Furuno, and Y. Korai, *J. Mat. Sci.* 23, 298 (1988).
87. I. Mochida, T. Oyama, and Y. Korai, *Carbon 25*, 273 (1987).

88. A. Y. Nesumi, T. Oyama, Y. Todo, A. Azuma, I. Mochida, and Y. Korai, *Ind. Eng. Chem. Res. 29* (1990).

88a. I. Mochida, Y. Q. Fei, Y. Korai, K. Fujimoto, and R. Yamashita, *Carbon 27*, 375 (1989).

89. Y. Q. Fei, K. Sakanishi, Y. N. Sun, R. Yamashita, and I. Mochida, *Fuel 69*, 261 (1990).

90. I. Mochida, Y. Todo, T. Oyama, and Y. Korai, *Ext. Abstracts 19th Bien. Conf. on Carbon*, University Park, PA, 1989, p. 198.

91. I. Mochida, Y. Korai, T. Oyama, Y. Nesumi, and Y. Todo, *Carbon 27*, 359 (1989).

92. I. Mochida, Y. Todo, T. Oyama, and Y. Korai, *Ext. Abstracts 19th Bien. Conf. on Carbon*, Penn State Univ., PA, 1989, p. 196.

93. I. Mochida, Y. Todo, T. Oyama, and Y. Korai, *J. Petroleum Inst., Japan*, to be published.

94. I. Mochida, Y. Q. Fei, and Y. Korai, *Fuel 69*, 667 (1990) .

95. I. Michida, Y. Tomari, Y. Iwanaga, K. Maeda, and K. Takeshita, *J. Fuel Soc. Japan 54*, 994 (1975).

96. M. Marsh, I. Mochida, E. Scott, and J. Sherlock, *Fuel 59*, 517 (1980).

97. I. Mochida, K. Amamoto, K. Maeda, K. Takeshita, and H. Marsh, *Fuel 58*, 482 (1979).

98. I. Mochida, I. Ueno, K. Sakanishi, Y. Korai, and H. Fujitsu, *J. Petrol. Inst. 30*, 31 (1987).

99. Y. Horita, T. Oishi, M. Nakamura, and H. Tanzi, *J. Petrol. Inst. 30*, 101 (1987).

100. M. Kimura, Y. Sanada, and H. Honda, *J. Fuel Soc. Japan 49*, 752 (1970).

101. I. Mochida, T. Ando, K. Maeda, H. Fujitsu, and K. Takeshita, *Carbon 18*, 131 (1980).

102. M. S. Spencer and T. V. Whittam, in *Catalysis*, Vol. 3 (C. Kemball and D. A. Dowden, ed.), Chemical Society, London, 1980, p. 189.

103. I. Mochida and K. Takeshita, *J. Petrol. Inst. Japan 20*, 183 (1977).

104. I. Mochida, K. Shimizu, and Y. Korai, *Carbon 26*, 843 (1988).

105. I. Mochida, K. Sakanishi, T. Oishi, Y. Korai, and H. Fujitsu, *Proceedings of International Conference on Science 63* (1985).

106. I. Mochida, T. Tahara, K. Maeda, H. Fujitsu, K. Takeshita, *J. Petrol. Inst. Japan 20*, 1025 (1977) and *23*, 21 (1980).

107. A. Karami, *J. Petrol. Inst. Japan 16*, 366 (1973).

108. M. Kakuta, N. Tsuchiya, and M. Kohriki, *Tanso 85*, 55 (1976).

109. I. Mochida, K. Sakanishi, Y. Korai, and H. Fujitsu, *Chem. Lett.*, 909 (1985).
110. K. Sakanish, X. Z. Zhao, H. Fujitsu, and I. Mochida, *Fuel Process. Technol. 18*, 71 (1988).
111. I. Mochida, K. Sakanishi, Y. Korai, and H. Fujitsu, *Fuel 65*, 1090 (1986).
112. I. Mochida, T. Oyama, and K. Fujimoto, *Tanso 131*, 187 (1987).
113. I. Mochida and H. Marsh, *Ext. Abstracts 17th Bien. Conf. on Carbon*, Lexington, KY, (1985), p. 276.
114. I. Mochida, R. Ohtsubo, K. Takeshita, and H. Marsh, *Carbon 16*, 25 (1980).
115. F. Carmona, P. Delhaes, G. Keryer, and J. P. Manceau, *Solid State Commun. 14*, 1183 (1974).
116. B. T. Kelly, *Physics of Graphite*, Applied Science Publishers, 1981, p. 21.
117. K. Fujimoto, M. Yamada, M. Iwasa, and H. Nigino, *High Temp.-High Press. 16*, 669 (1984).
118. I. Letizia and N. H. Wagner, *Ext. Abstracts 16th Bien. Conf. on Carbon*, San Diego, 1983, p. 593.
119. I. Letizia, *High Temp.-High Press. 9*, 291 (1977).
120. M. Kakuta, H. Tanaka, J. Sato and K. Noguchi, *Carbon 19*, 347 (1981).
121. I. Lindhout, R. S. Downing, and F. J. A. Geiger, *Ext. Abstracts Carbon '88*, Newcastle upon Tyne, UK, 1988, p. 380.
122. M. P. Whittaker and L. I. Grindstaff, *Carbon 7*, 615 (1969).
123. I. Mochida, T. Furuno, H. Fujitsu, T. Oyama, and K. Fujimoto, *Fuel 67*, 678 (1988).

4

Interfacial Chemistry and Electrochemistry of Carbon Surfaces

Carlos A. Leon y Leon D.* and Ljubisa R. Radovic

The Pennsylvania State University, University Park, Pennsylvania

Current affiliation: Quantachrome Corporation, Syosset, New York

I. INTRODUCTION

Carbon materials in general, and carbon reactions in particular, have repeatedly proven to be an inexhaustible source for scientific research and technological development. New discoveries and improved understanding lead to an ever-increasing number of potential applications even when the relevant industries show signs of maturity. Combustion of carbon [1,2] on our planet has been studied for centuries. Carbon–carbon composite materials [3,4] are assisting us in the exploration of space surrounding our planet. And the pursuit of knowledge of the composition of space between planets and stars has given us the fullerenes [5,6], molecules that have recently aroused the admiration of scientists and the public alike.

At the heart of carbon reactions and the key to the usefulness of many carbon materials is the interfacial chemistry of carbon surfaces. Yet, even though the field of carbon reactions has been reviewed frequently [7–10], carbon surface chemistry has not received the attention that it deserves. A summary was provided in this series by McKee and Mimeault [11] in the context of a more general review of the surface properties of carbon fibers. More recent reviews can be found dispersed in several chapters of the books by Kinoshita [12] and Bansal et al. [13]. Brief overviews have also been recently published by Boehm [14] and Tarasevich and Khrushcheva [15]. For additional information, one must consult the early pioneering works by Donnet and Voet [16], Randin [17], Mattson and Mark [18], Rivin [19], Puri [20], Deviney [21], Boehm [22], Garten and co-workers [23] and Studebaker [24], among others. These outstanding contributions provide fascinating reading to those interested in understanding carbon surface chemistry. However, it soon becomes clear to all readers that, because the available literature consists of contributions by different research disciplines, it is regrettably scarce in unifying concepts. The objective of our contribution is to fill this gap by attempting to provide a more unified, comprehensive, and up-to-date review of carbon surface properties and to illustrate their importance in some of the most fecund application of these materials, i.e., as adsorbents, catalyst supports, catalysts, and reactants.

Since the chemistry and physics of carbon materials are closely intertwined, it is appropriate to begin the discussion with a summary of the least controversial aspects of carbon structure and properties of carbon surfaces (Section II). Sections III–VI offer reviews of the current state of the art from the perspectives of physical chemistry, solid-state chemistry, analytical chemistry, and electrochemistry. In Sections VII and VIII, we aim to illustrate—using our own most recent work as well as that of others—how the understanding of surface chemistry of carbons can help to understand, and indeed optimize, their use in catalysis and gasification.

II. ESTABLISHED FACTS ABOUT CARBON SURFACES

Because of their different sources, degrees of decomposition, and purities, solid carbons make up a very long list of materials, which includes items as diverse as natural or synthetic graphite, coke, soot, carbon blacks, activated carbons, and charcoal. In spite of their diversity, solid carbons share three distinguishing features:

A high content of elemental sp^2-hybridized carbon exhibiting at least two-dimensional order
The ease with which they combine with other elements (e.g., oxygen) to form a variety of surface complexes
The transient nature of their surface properties

The latter two features have important consequences for the utilization of these materials. On the positive side, the tremendous flexibility of solid carbons provides the possibility to produce materials with optimum surface chemical, electrochemical, and physical properties for selected applications. On the other hand, the final properties (in particular the chemistry and the electrochemistry) of solid carbons are invariably difficult to predict, to evaluate, and, oftentimes, to reproduce. For these reasons solid carbons have generally found wide acceptance mostly in areas in which they have proven, even without their optimization, to be far superior to other materials. Evidently, much can be gained by striving to develop a better understanding of the surface properties of solid carbons, and to prove them of significant value for a wider range of applications of commercial impact.

In trying to develop a firm basis for rationalizing the chemical and electrochemical responses by carbon surfaces, it is instructive to review all pertinent approaches reported in the literature. The accumulated knowledge about the properties of solid carbons relies heavily on the interpretations given over the years to data generated using a wide variety of experimental techniques. Unfortunately, when applied to solid carbons, many physical and virtually all chemical characterization tools fail to provide direct, unambiguous results. Nevertheless, all techniques that uncover differences among samples are important, because their underlying theories and inferences provide (i) a much needed frame of reference and (ii) a basis for critical, comparative assessment.

Based on such considerations, it has been possible to generate a sufficiently coherent picture of the physical (structural) characteristics of solid carbons [25,26]. Accordingly, solid carbons are taken to consist mainly of carbon atoms grouped into layers of fused aromatic rings exhibiting a certain degree of planarity (or two-dimensional order). This planarity can be extended from a few rings (as in nongraphitic carbons) to thousands of them (as in graphitic carbons). In either case, the individual layers are likely to contain imperfections (e.g., twists, non-aromatic links, and carbon vacancies), in spite of which they tend to stack more or less on top of each other, and are held in position by weak van der Waals forces (ca. 3 kcal/C atom [27], versus > 70 kcal/C atom for carbon–carbon bonds). Heating to > 2800 K in nonreactive environments may (for graphitizable carbons) or may not (for nongraphitizable ones) smooth the imperfections and improve the stacking order, so as to make the structure of the material resemble, more or less, that of graphite. Less graphitic carbons have smaller, less well-stacked layers (i.e., smaller crystallites). Nongraphitic carbons (e.g., activated carbons and carbon blacks) are dubbed microcrystalline because some evidence suggests that in any given particle none of its many imperfect crystallites exceeds ca. 5 nm in size [12,13,16].

These structural arrangements are responsible for some remarkable chemical, electronic, and electrical properties of carbons and graphite, which clearly set them apart from other materials, as can be gathered from, e.g., a quick comparison with the surface chemical properties of inorganic solids. For instance, metal oxide surfaces consist mainly of oxygen atoms, hydroxyl groups, and a few exposed (coordinatively unsaturated) metal atoms [22,28,29]. It is primarily the surface hydroxyl groups that determine the chemistry (acid–base character) and the reactivity (e.g., in ion or ligand exchange) of these surfaces [22]. The observed chemical heterogeneity of metal oxide surfaces is related to the effect of the local environment of each hydroxyl group [28]. Coordination of the hydroxyl groups with one ("linear"), two ("bridged"), three, or more lattice metal atoms can yield a variety of sites having different properties. Furthermore, neighboring sites can be affected by hydrogen bonding between them. The net result is that metal oxide surfaces of the form $MO_x(OH)_y$ present an assortment of surface hydroxyl groups ranging, in general, from Brønsted acids (high x/y ratio) to Brønsted bases (low x/y ratio). In contrast, carbon surfaces can contain not one but, at the very least, five markedly different type of surface groups (i.e., carboxylic-, lactonic-, phenolic-, carbonyl-, and etheric-type) [22,28]. Even though their relative distribution is dependent upon the type of carbon considered and its pretreatment [30], the properties of each individual group are again determined by its specific local environment. It follows that, even without considering the Lewis acid–base properties of these materials, the surface chemistry of solid carbons must be much more versatile than that of metal oxides.

Before proceeding, it is important to address the need (or lack thereof) for a conventional distinction between surface chemistry and electrochemistry. All chemistry is, of course, electrical by nature. Hence, in a sense, all chemistry is "electrochemistry." However, the latter term is specifically reserved for processes that involve a transfer of electrons between the participating species. Since electron transfer processes entail a flow of current, their details can be monitored and rationalized using electrical principles. On the other hand, since these processes are invariably associated with chemical (oxidation–reduction) reactions, their details must also abide by chemical principles. Most textbooks treat electrochemistry as a branch of chemistry. However, a quick perusal of the relevant literature shows that, at least as far as carbon surfaces are concerned, both areas have grown apart over the years. In fact, their estrangement has been very detrimental to the development of applications for carbon surfaces [30]. This factor undoubtedly contributes to the widespread notion that carbon surfaces are chemically too complex to be fully characterized.

Attempting to understand the surface chemistry of solid carbons can certainly prove to be a monumental task. This has been a challenge for scientists during at least the last two centuries. (The first recorded account of the existence of surface oxides on carbon surfaces was made at least as far back as in 1814 [31], and not in 1863, as suggested in an often cited reference [20].) During this time an immense number of experimental observations have been made, and it is not surprising that there are a variety of proposals found in the literature to explain the chemical behavior of carbon surfaces. Indeed, very valuable fundamental contributions have been made by researchers from virtually all branches of chemistry. Nowadays, advocates of the following four schools of thought can be distinguished:

Electrochemists, who tend to view carbons as electrodes whose surfaces contain sites where redox reactions can take place [15,17,32–34]

Analytical chemists, who prefer to regard solid carbons as macromolecules whose surfaces are capable of holding all kinds of functional groups (analogous to those of small organic molecules) [12–14,16, 18–20,22–24,35–37]

Solid-state chemists, for whom solid carbons, despite their disordered lattices, should be viewed as semiconductors with π bands, holes, and free radicals at edges and defects of graphitic crystallites [38–40]

Physical chemists, for whom carbon surfaces are made up of fairly unreactive basal planes along with edges decorated primarily by surface oxides of the general form $C_z(CO)_x(CO_2)_y$ [41–45].

Taken individually, each line of reasoning can be exploited for specific applications. Hence, it is common for researchers to adopt the line of reasoning that is most consistent with their background and to overlook (all) others. This pervasive habit is reflected in the contents of most literature reviews, which very seldom touch on all aspects of carbon surface chemistry. If all accumulated knowledge of carbon surface chemistry is to be placed in a proper perspective, it is necessary to realize that no single discipline can claim to successfully explain all surface chemical properties of solid carbons. (In fact, all are necessary and complementary.) Also, no clear links have been established among the different disciplines previously mentioned, with electrochemistry in particular being notoriously distant from the rest. In Sections III–VI the hypotheses put forward by the aforementioned groups are critically evaluated and contrasted. Their experimental techniques, summarized in

TABLE 1 Selected Techniques for the Chemical Characterization of Solid Carbons

Technique	Main Target	Reference(s)
Voltammetry	Redox processes	[12,46]
Potentiometry	Surface reactions	[13,47]
Polarography	Redox processes	[13,48]
Coulometry	Redox processes	[48]
Electrokinetics	Surface charge	[49,50]
	Isoelectric point	[49,50]
Acid-base titrations	Functional group type/number	[36,51,52]
Mass titrations	Point of zero charge	[53,54]
Temperature-programmed desorption	Functional group type/number	[12,19,55,56]
	Functional group energetics	[57,58]
Temperature-programmed oxidation	Reactivity	[59]
X-ray photoelectron spectroscopy	Functional group type/(number)	[60–62]
(Fourier Transform) infrared spectroscopy	Functional group type/(number)	[63–65]
Electron spin resonance	Free electrons/radicals	[38,39]
Nuclear magnetic resonance	Degree of aromaticity	[27,39]
Magnetic properties	Electron conduction mode	[38,39,66]
Wetting by liquids	Contact angles	[67,68]
Contact potential	Work function	[69]
Calorimetry	Sorption energetics	[70–72]
	Surface polarity	[73]
Other chemical reactions	Functional group type/number	[27,35]
Electrochemical reactions	Redox properties	[23]

Table 1, are analyzed only in the context of what they reveal about carbon surfaces. Special emphasis is placed on attempts to develop unifying concepts, in the hope that these will lead to a more coherent picture of the chemistry of carbon surfaces.

At the present time, the following are generally agreed upon as being well-known facts about carbon surfaces:

They can form chemical or physical bonds with a wider spectrum of substances than any other material [20,21,23,24,36].

They can accommodate Brønsted acid sites [20,36], Lewis basic sites [24], and sites capable of taking up or giving up single electrons (i.e., oxidizing or reducing sites, respectively) [23].

By most accounts, the surface coverage by any or all types of sites seems not to be directly related to the total surface area of the carbons. For example, a value of 1 mmol/g of sites, typical of both high- and low-surface-area carbons, would correspond to ca. 100 m^2/g of their surface.

Even very small contents of heteroatoms (e.g., oxygen, hydrogen, nitrogen, sulfur) can exert a significant influence on the chemical properties of any given carbon.

All heteroatoms are held on tenaciously to the surface and can only be removed at the expense of some carbon (e.g., oxygen adsorbed at temperatures as low as 233 K being primarily released, upon heating, as CO and CO_2 [13]).

Oxygen is invariably the most common heteroatom found on carbon surfaces, a fact that explains the large number of studies devoted to oxygen complexes [20,36].

The more oxygen they hold, the more carbon surfaces behave like those of oxides, particularly in terms of their adsorption properties (the term graphite oxide being reserved for a specially prepared nongraphitic carbon [74]).

Their properties change during storage [18].

A complete model for the chemical properties of carbon surfaces must make allowances at least for these facts. No single model or theory can do so at the present time. However, a combination of theories may succeed, as long as the theories invoked are consistent with each other. All theories derived from studies of carbon surfaces are usually developed by making some simplifying assumptions. The most relevant assumptions and contributions made by each discipline to the study of carbon surfaces are critically evaluated in the following sections. The only inherent assumption that shall be made throughout this review is that all solid carbons, regardless of their source (provided that it is clear that inorganic matter plays at best a minor role), are capable of displaying the same types of chemical features. In other words, solid carbons shall be considered unique *as a whole* rather than individually.

III. CONTRIBUTION BY PHYSICAL CHEMISTRY TO THE STUDY OF CARBON SURFACES

One notion common to all disciplines is that clean carbon surfaces are made up of (at least) two chemically different kinds of sites (i.e., basal and edge carbon atoms). The validity of this assumption was confirmed for highly pure graphite, by noting that the reactivity of edge sites toward oxygen is over one order of magnitude higher than that of basal sites [75]. Graphite has very low edge-to-basal site ratios, but its edges are very reactive, since they contain single unpaired electrons [76]. Some of these edge sites are occupied by strongly bound heteroatoms (most commonly hydrogen or oxygen). As the structural order of a solid carbon decreases, its edge-to-basal site ratio increases, and so does the carbon's overall reactivity [77]. Three decades ago it was proposed that edge sites on graphitic carbons could be titrated via low-temperature (e.g., 573 K) oxygen chemisorption [43]. The concept of an active surface area (ASA) [43–45,78–80] was then introduced in an attempt to quantify the fact that not all the accessible surface of solid carbons is equally reactive. (An adsorption stoichiometry of one oxygen atom per surface carbon atom, and an oxygen coverage of 0.083 nm^2/atom, were assumed in the calculations [43].) The ASA postulate proved to be very useful as a correlator of the gasification reactivity of graphitic carbons, but its extension to highly disordered carbons met with limited success [81–83]. Nevertheless, it has been pointed out recently that, despite its limitations, the ASA concept provides values that should be at least qualitatively related to the content and the chemical response (in polar solvents) of oxygen functional groups on carbon [45]. Such an expectation is based on the belief that both these parameters are also related to the concentration of edge sites.

Unfortunately, pertinent investigations suggest that the relationship between ASA and surface chemistry is not straightforward [30,45]. Its complexity may be related to a number of aspects, e.g., the assumed adsorption stoichiometry [62,84], the assumed oxygen coverage (which is an upper limit for graphite and a low value for microcrystalline carbons [85]), the arbitrary choice of experimental conditions (O_2 exposure time, pressure, and temperature) for ASA determination, and the fact that to measure ASA values it is customary to remove the initial oxygen functional groups by thermal treatment at elevated temperatures (e.g., 1223 K). The higher the initial functional group content of a solid carbon, the greater will be the loss of carbon atoms (carried away as CO and CO_2) prior to ASA measurements. For a typical functional group content of 1 mmol/g, the desorption of between 1.6 and 3.2 wt% oxygen as carbon oxides would carry along anywhere between 0.6 and 2.4% of carbon. Carbon losses of this magnitude can significantly affect the chemistry and the physics of porous carbons. Indeed, the stripping of carbon oxides generates a totally new surface, which may or may not resemble that existing prior to the desorption of functional groups. High pyrolysis temperatures can also propitiate thermal annealing (particularly for high initial oxygen contents [19]), which could in turn change the surface properties of carbon considerably [19,86]. These and other possible sources of interference [83] preclude the direct use of the ASA concept as a fundamental correlator of carbon surface chemical properties. (Its use as a correlator of carbon reactivity is briefly reviewed in Section VIII.)

On the other hand, a direct thermal treatment of the samples, as in temperature-programmed desorption (TPD) or temperature-programmed oxidation (TPO), can provide a wealth of information about carbon surfaces, if it can be shown that the contribution to the thermal responses by secondary reactions [58] is small. Inferences from thermal treatments and their quantitative relation to the surface chemistry of porous and nonporous carbons are discussed at length elsewhere [30] and summarized in Sections VII and VIII.

Additional contributions by physical chemistry have been made through the application of spectroscopic techniques such as (Fourier transform) infrared spectroscopy ((FT)IR) or x-ray photoelectron spectroscopy (XPS) and by drawing inferences from measured physical properties (e.g., contact angle, work function, and heat of wetting). All these techniques provide very useful *qualitative* information about carbon surfaces (see Table 1). *Quantitative* agreement among these and other (e.g., titration or TPD) methods is also occasionally reported [12,87,88]. However, the validity of their quantification is, at present, open to debate [39]. For instance, infrared (IR) experiments yield energy absorption peaks at specific wavelengths, which can be identified by comparing them with the wavelengths of known compounds [12,65]. (Band assignments to groups such as carboxyl, phenolic, and aromatic and aliphatic hydrogen have thus been reported [12].) Even though the intensities of the IR absorption peaks are presumed to be linearly related to the concentration of the species giving rise to them, the necessary conversion factors cannot be unequivocally established for complex materials like carbons [65] or coals [89].

Similarly, XPS experiments yield photoelectron peaks at specific irradiation energies, which can be related to the binding energies of core electrons ejected from atoms located on the external surface (typically 1–5 nm thick) of a sample [12,61]. These binding energies and their shifts can in turn be related to those of known compounds [61,90]. Upward shifts in binding energies result from a decrease in electron density around the nucleus in question (i.e., from oxidation), whereas downward shifts result from its reduction. Linear correlations between these shifts and the charge on given atoms can be expected [91] and have been reported for carbon–oxygen complexes (see Section III) [61,88]. In this manner singly, doubly, and singly-plus-doubly (or triply) bonded carbon–oxygen groups have been associated with phenolic, carbonyl, and carboxyl complexes, respectively [92].

However, an unambiguous interpretation of XPS spectra is difficult. Problems arise because, e.g., (i) the external (probed) and the internal composition of a solid carbon sample may differ, and they often do [30,52]; (ii) in porous carbons external voids can constitute a portion of the probed space and artificially lower the measured concentration of surface species, especially those close to, but not necessarily at, the gas–solid interface (e.g., carbon); (iii) the nature of the surfaces at ultrahigh vacuum (ca. 10^{-9} torr) conditions used for XPS experiments may differ from that at normal conditions, owing to, e.g., surface restructuring in the latter case [93]; (iv) the peaks are affected by well-known charging and relaxation effects [91b]; and (v) the deconvolution of lumped C_{1s}, O_{1s}, or N_{1s} peaks into individual components [13,90,91,94] is far from being straightforward at present [30].

In the case of contact angle, surface work function, heat of wetting by liquids, and other such physical measurements (Table 1), current approaches seldom account for the heterogeneous nature of carbon surfaces. Consequently, surfaces (or even particles) that present a distribution of site energies are assigned experimentally derived mean contact angles, work functions, heats of wetting, etc., which in fact account only for the response of a fraction of the available surface sites (i.e., that found on or close to the external surface of some or all particles). Typical values reported for the previously cited variables are in the range 0–110° [67,68], 4.5–5.5 eV [69], and 1–20 cal/g [12,18], respectively. Such information may be satisfactory for applications of carbons that depend mostly on the properties of their external—geometric—surfaces (e.g., rubber reinforcement by carbon blacks [16,21]). In such cases, inferences may be drawn regarding the overall behavior of carbon surfaces. For example, both the surface work function (Section VI.B) and the contact angle of water on various carbons have been reported to be pH-dependent [68,69]. (These observations are better understood in terms of contributions by analytical chemistry, as discussed later.) Nevertheless, other applications (e.g., catalyst manufacture) would greatly benefit from the introduction of statistical approaches to determine these variables [68]. In general, it is more sound to define a heterogeneous system in terms of two variables (i.e., a mean and a standard deviation) than in terms of only one (a mean in the foregoing cases). In the absence of such approaches, the interpretation of these properties remains as controversial as, for example, the application of the BET theory to the physical description of heterogeneous surfaces [95,96].

IV. CONTRIBUTION BY SOLID-STATE CHEMISTRY TO THE STUDY OF CARBON SURFACES

Describing (most) solid carbons as semiconductors is of value in that it provides an explanation for a number of electronic, electrical, and thermal properties of carbon surfaces [38–40]. The conductivity of solid carbons arises from two types of charge carriers: π electrons and holes in the π band of polycondensed aromatic ("graphene") layers. In both cases, charge can flow (i) along individual graphene layers with much more ease than flowing perpendicular to them and (ii) only perpendicularly between different layers [27]. Most solid carbons, being much less ordered than graphite, have very ill-defined "bands." Strictly speaking, solid carbons should not be considered semiconductors, even if they behave as such, because their enhanced conductivity at higher temperatures is due to a structural rearrangement (graphitization), rather than to the thermal excitation of valence electrons. Nevertheless, it is useful to consider the inferences drawn from semiconductor evaluation techniques regarding their application to the study of solid carbons.

Electrical conductivity measurements, despite their limitations when applied to porous powdered solids [27], provide some valuable information regarding carbon surfaces [12,39]. The conductivity of solid carbons changes as a function of heat treatment; generally, it increases rapidly below ca. 973 K and continues to increase at a much slower pace above ca. 973 K [38,39]. The temperature range in which the conductivity of solid carbons begins to level off (873–973 K) coincides with that in which solid carbons lose their acidic properties [19,20,97,98]. Most surface properties of solid carbons are closely related to the heteroatoms, which are generally lost upon heat treatment as H_2O (< 623 K), CO_2 (< 923 K, with a few exceptions), CO (< 1473 K), H_2 and CH_4 (> 1223 K), and nitrogen and sulfur forms (> 2073 K) [12,19,20,99]. Therefore, the loss of acidic functionalities (primarily as CO_2) approximately coincides with the leveling off of the electrical conductivity of solid carbons. Both factors are presumably related by electron delocalization effects. Electron delocalization, which favors electrical conduction, is enhanced by the loss of strong electron-withdrawing (CO_2-evolving) groups. Upon the complete loss of CO_2-evolving groups, the conductivity would tend to level off, as observed in practice [39]. Since the conductivity continues to increase slightly at temperatures above 973 K, there is probably an additional contribution to electron delocalization due to graphene layer growth following thermal annealing, or "densification" [86]. These possibilities are better assessed using more refined solid-state techniques such as electron spin resonance (ESR) [12,39].

Electron spin resonance is a technique that is sensitive to the concentration of unpaired electrons, such as delocalized π electrons and localized—unpaired σ, quinonoid, etc.—free radicals [12]. This technique allows one to monitor specific changes in electronic properties of solid carbons (e.g., as they are heated). For instance, ESR indicates that the loss of heteroatoms below 973 K generates stable free radicals on the surfaces (edge sites) of solid carbons [38,40]. These immobile free radicals are quite stable and at low concentrations (i.e., at low desorption temperatures) can be expected to be rather distant from each other [39]. As the temperature increases, the loss of heteroatoms generates more and more radicals, until in the range 873–973 K they become so numerous that they start recombining amongst themselves [39]. For many solid carbons, the maximum concentration of free radicals (detected in the range 873–973 K) is generally estimated to be $\leq$ 0.3 mmol/g [16,21,22,27,39,40,100,101], i.e., only a fraction of their typical functional group content detected by TPD [30] or even by acid–base titrations (see Section V). For instance, solid carbons with surface areas > 500 m^2/g have been reported to contain maximum free radical concentrations within about 1–10% of their functional group content determined by neutralization with NaOH (see Section V) [100,101]. In these cases, the concentration of unpaired electrons was found to be proportional to the volatile matter content (following heat treatment and/or oxidation) of the carbons. This was taken as an indication that one or more types of oxygen functional groups could act as electron acceptors [100]. Indeed, the activity of carbons as oxidation catalysts (i.e., electron acceptors) appears to be related to the concentration of edge sites [102–105], where oxygen functional groups are located. The suggestion was made that quinonoid complexes could account for ESR responses [22,100] and for other properties [16,21,23] of solid carbons. A simplified scheme for the proposed mechanism is shown in Fig. 1.

Quinonoid complexes, often called quinones for simplicity, consist of a couple of carbonyl groups placed on the edge of a common graphene layer (structure I). Resonance considerations permit the stabilization of a number of radicals (structure II) in equilibrium with structure I. The stability (and possibly the number) of the resulting radicals is proportional to the size of their interconnecting graphene layer [106,107].

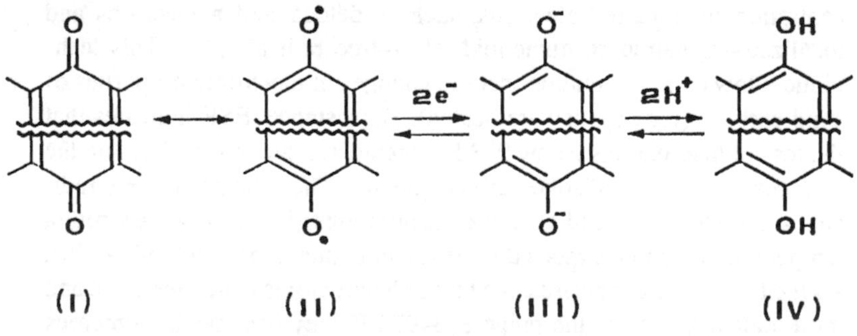

(I)   (II)   (III)   (IV)

FIG. 1 Electron transfer activity of quinonoid (quinone) sites. Two carbonyl groups located at edges of a common aromatic layer (I) can act in concert to stabilize radicals (II) via resonance. The radicals can accept electrons and become anions (III). The anions can transfer electrons back and become radicals (II), or they can interact with cations in solution. Reversible proton transfer leads to phenolic ("hydroquinone") sites (IV).

These radicals may take up electrons and become anions (structure III). The anions may further bind protons in solution, as shown in structure IV. This behavior resembles very much that encountered on quinhydrone electrodes, for which the reaction $C_6H_4O_2 + 2H^+ + 2e^- = C_6H_4(OH)_2$ is well documented [108,109]. The conversion of radicals to anions (II to III) need not be complete [110]; moreover, it can be expected that the depletion of radicals by electron addition (II to III) is compensated by the generation of new radicals (I to II) to maintain their equilibrium with structure I. The scheme shown in Fig. 1 is consistent with IR evidence [65] and with a number of electrochemical observations (see Section VI), which confirm that electron-transfer processes do take place on carbon surfaces.

On the other hand, the suggestion that other (oxygen [23,100] or nitrogen [90,105]) functional groups besides quinonoid complexes could be similarly involved in electron transfer processes has only been indirectly substantiated (by electrochemical methods; see Section VI.A). However, it can be presumed that the only requirement for a heteroatom bearing an unshared pair of electrons (e.g., oxygen, nitrogen, sulfur, and halogens) to act as shown in Fig. 1 would be its association with a substrate (e.g., an aromatic network) containing free radicals stabilized by resonance [106,107]. Schemes such as that shown in Fig. 1 are essential in order to make a full account of the chemical and electrochemical versatility of solid carbons, as will become apparent in subsequent sections.

The fact that the concentration of unpaired electrons detected by ESR is rather low was attributed to quinonoid complexes (or other oxygen-containing analogues) being located on very large graphene layers [100]. However, it was also hypothesized that unpaired electrons could simply be transferring a fraction of their charge (by partial delocalization in an area on or near the surface), and thus escape detection by ESR. In fact, stable radicals are usually in equilibrium with other resonance-stabilized forms [106,107], so only a fraction of available quinone groups would be expected to generate radicals stable enough to allow their detection by ESR. Surprisingly, no electron donor sites were thought to be present [100a], even though it was later found that the same materials were capable of strongly adsorbing some molecular oxygen, which could be subsequently reduced electrochemically (see Section VI) [100b]. This oxygen is presumably held as superoxide ions, O_2^-, on carbon surfaces (see Section VI.A) [23,65,90]. If so, the incomplete removal of adsorbed oxygen prior to ESR measurements could be partly responsible for the observed p-type semiconductor properties of many solid carbons, particularly following low-temperature heat treatments (see following discussion) [39].

Results obtained by ESR also indicate that, upon heat treatment above ca. 973 K, the elimination (by recombination) of free radicals is paralleled by a rise in the concentration of free electrons, believed to be trapped in carbon vacancies (as spin pairs) in graphene layers [27,38,39]. The concentration (or the mobility [39]) of free electrons also goes through a maximum with increasing temperature, and decreases significantly above 1673 K (i.e., once the band gap is virtually closed) [39]. The occurrence of an excess of free electrons after heat treatment at these temperatures (973–1673 K) is also consistent with the p-type semiconductor properties of many solid carbons (discussed below). In all cases, ESR studies make clear that the electronic properties of solid carbons at room temperature depend markedly on the heat pretreatment to which they are subjected.

An alternative way to monitor the above changes conveniently is by means of the Hall effect [39]. The Hall coefficient (H) gives the predominant sign and concentration of charge carriers in solid carbons. If the majority of charge carriers are π electrons, then $H < 0$ and the material is dubbed n-type semiconductor. Conversely, if the majority of charge carriers are holes in the π conduction band, then $H > 0$ and the material is dubbed p-type semiconductor. The conduction mode of a given carbon is a strong function of its final pretreatment temperature [38–40]. An interesting situation is found in the case of solid carbons; at room temperature, $H < 0$ for graphite (implying conduction by π electrons) and $H > 0$ for disordered, nongraphitic carbons (implying conduction by holes). As the nongraphitic carbons are treated, they may or may not undergo graphitization. If they do not, H is likely to remain positive; if they do, H has been found to decrease with increasing temperature and even become slightly negative prior to the onset of significant two-dimensional layer growth (at ca. 1273 K) [39]. Up to this stage, the band gap amounts to < 10 kcal/mol [40]. At somewhat higher temperatures (e.g., 1473 K) the band gap closes [38,40], and H becomes positive again. Only at the onset of significant three-dimensional ordering (e.g., 2073 K) does H begin to decrease, until at graphitization temperatures (> 2773 K) its value becomes negative. These observations are important because they indicate that in most solid carbons one is to expect conduction by holes to predominate. Hence, solid carbons must be regarded predominantly as p-type semiconductors [40], unless they had been subjected to very high (graphitization) temperatures. (In some cases, mild heat treatment in the range 973–1273 K has been reported to produce net n-type semiconductors [39].) In terms of surface properties, these are closely related to the predominant bulk conduction mode of a given carbon. Accordingly, in p-type semiconductors, holes are generated in the bulk by the net migration of electrons toward the surface, the converse being true of n-type semiconductors [111]. This suggests that the surfaces of individual crystallites (should they exist as such) in nongraphitic carbons would be enriched in π electrons at the expense of those in the bulk. These extra π electrons are believed to migrate toward the edge sites (where they form spin pairs with otherwise unpaired σ electrons), thereby leaving holes in the π band of their corresponding graphene layer [112]. In all likelihood, these extra π electrons would confer the exposed basal planes an electron-donor (reducing) character, while the unpaired sigma electrons at edge sites would remain as electron acceptors. Most of these electrons would temporarily be placed as spin pairs on crystallite edges [39]. These inferences are consistent with evidence indicating the ease of formation of O_2^- ions on carbon surfaces (mentioned previously) [23,65,90] and with the general tendency of solid carbons to form bonds with heteroatoms in proportion to the concentration of their edge sites [12,20]. They also help to explain several chemical and electrochemical properties of carbon surfaces, as discussed in later sections.

Unique features of carbon materials can also be determined using solid-state nuclear magnetic resonance (NMR) [27,39]. Even though the principles of NMR and ESR techniques are comparable, the application of NMR to solid carbons is not simple. Complications arise because, among other factors [39], NMR cannot detect ^{12}C. However, ^{1}H, ^{13}C, and presumably [37] ^{17}O are responsive to NMR. Both ^{1}H and ^{13}C NMR can detect large enough amounts (i.e., about 10–30 and 0.7–1 atom %, respectively) of the total organic matter content of typical solid carbons to allow their characterization. (In contrast, ^{17}O is present only in ppm atomic levels.) In the case of ^{13}C, since only ca. 1.1% of the available carbon can be probed, the optimistic assumption must be made that this portion is representative of the whole sample. Recent refinements in the technique [39] allow the deconvolution of NMR peaks into components arising from different local environments surrounding the atoms probed. However, until all complications are resolved [39], the information obtainable from NMR spectra of solid carbons is limited to an estimation of their aromaticity, i.e., the fraction of carbon atoms present in aromatic rings. This also gives an indication of their nonaromatic (i.e., aliphatic or alicyclic) carbon content. Typical solid carbons are thus estimated to contain 1–10% of their elemental carbon atoms bound in nonaromatic fashion [27,39]. The chemical properties of nonaromatic and aromatic portions of solid carbons are bound to be different; two clear examples of their differences are found in their gasification reactivity—the latter, more "ordered" and resonance-stabilized being gasified less readily than the former—and their ease of wetting by liquids, the former being, e.g., more hydrophobic than the latter [68].

Finally (and paradoxically), a much overlooked factor regarding solid carbons is that they are black materials. This means that they readily absorb heat and light. Typical heat capacities of solid carbons range between 0.2 and 0.4 cal/g/K at room temperature [27]. These values are high enough to allow an increase in the temperature of bulk amounts of solid carbons (in particular moist samples) stored in spaces where heat dissipation is limited. Also, as discussed previously, light absorption in the IR region introduces enough energy to cause the vibration of covalently bonded atoms, thereby giving rise to identifiable peaks. The amount of energy E (in kcal/mol) introduced by light of wavelength λ (in nm) can be estimated from the relationship $E = 28,591/\lambda$ [113]. Accordingly, infrared ($\lambda > 760$ nm), visible ($700 < \lambda < 400$ nm), and ultraviolet ($\lambda < 350$ nm) light carry < 37, 41–72, and > 82 kcal/mol, respectively. In the latter two cases, enough energy is available to induce photochemical reactions (via, e.g., the homolytic cleavage of covalent bonds) [107,113]. (In this context, it should be noted that chlorophyll, which makes photosynthesis possible, is able to absorb visible radiation by virtue of its special conjugated ring system [114]; also, many ketones are known to be particularly susceptible to photochemical cleavage [107].) Therefore, the prolonged illumination of solid carbons by light may bring about marked changes in their surface properties, analogous to those leading to coal weathering [114] as far as surface complexes are concerned. This is especially true in the presence of oxygen and moisture [114,115]. Since little attention is often paid to the conditions of sample storage, this factor undoubtedly contributes to the transient (and sometimes irreproducible) nature of carbon surfaces.

V. CONTRIBUTION BY ANALYTICAL CHEMISTRY TO THE STUDY OF CARBON SURFACES

The views offered by analytical chemistry have dominated the minds of many carbon scientists for the past four decades. This is so mainly because such views are believed to provide the most detailed account of the surface chemistry of carbons at a molecular level. Many analytical procedures are available to identify the *functional groups* of unknown substances and estimate their content. All of them are based on chemical reactions, whose kinetics and thermodynamics are not always predictable. Therefore, it is customary (and wise) to support chemical evidence with results from other (e.g., physical chemistry) characterization approaches.

Functional groups are defined as groups of atoms in an organic compound that respond in distinct ways to different chemical reactants. These reactive groups can therefore be characterized by following systematic approaches [116]. Acceptable chemical procedures for the identification of functional groups range in the hundreds [12,13,16, 18–24,27,35,36,51,87,107,116], in part because of the wide variety of functionalities known to exist on simple organic molecules [116]. When applied to solid carbons, ambiguities in the results are unavoidable because of the following facts: (i) equilibration times, particularly for microporous carbons, are very long; (ii) grinding the particles to speed up the reaction(s) can alter their surface properties significantly, often leading to irreproducible results; (iii) the access of certain reactants to inner/smaller pores may be physically or chemically restricted (owing to their size, shape, electrostatic repulsion, etc.); (iv) the proximity of two or more functional groups alters the properties of each, by induction, chelation, and/or steric effects; (v) the nature of the carbon substrate also affects the properties of groups on its edges, by resonance and/or steric effects; (vi) the reactions are often accompanied by unwanted side reactions (in parallel and/or in series); and (vii) the reactions are rarely reversible and seldom go to completion. These facts explain why, excepting for their use in some specific applications [13,16,21,35,110], chemical characterization methods are not as popular with solid carbons as they could be. Nevertheless, analytical chemistry approaches allow the classification of different functional groups (and surfaces) according to their properties. Two useful classifications are those based on their acid–base and their oxidation–reduction properties. Even though some overlap between the two classifications is inevitable, for the purposes of this discussion acid–base properties will be those resulting primarily from ion (Brønsted) and/or electron pair (Lewis) interactions, whereas oxidation–reduction (redox) sites, to be dealt with in the section on electrochemistry (Section VI), will be those capable of engaging in one-electron transfer process. In the remainder of this section emphasis shall be placed on the elucidation of the acid–base nature of carbon surfaces.

The unique character of most functional groups stems from the fact that they are made up of heteroatoms whose electronegativities are different from those of carbon atoms. The more electronegative atom in a bond tends to acquire a negative partial charge, leaving the other atom with a positive partial charge. Charge separation makes the bond polar and influences its interactions with other bonds via, e.g., electron localization and dipole interactions. The extent of these effects is related to the actual polarity of the bond in question, which may range from covalent, due to full cloud overlap, to ionic, due to full cloud separation. Electronegativities for the major constituents of solid carbons (on the relative Pauling scale [107]) decrease as follows: $O(3.5) > N(3.0) > S(2.5) = C(2.5) > H (2.1)$. These imply that bond polarities can be expected to be much lower on C–S and on C–H bonds than on C–N and, especially, C–O bonds. Moreover, bond polarities are also influenced by the degree of electron cloud sharing of each atom with other surrounding atoms. Therefore, functional groups on carbon surfaces have bond polarities that are strongly influenced by the local nature of their substrates. It follows that functional groups on carbon surfaces differ from those of simple organic substances in that, even for the same group type (e.g., quinonoid, as in Fig. 1), *no two groups can be expected to behave exactly alike*. At best, they will form a continuum with a distribution of bond polarities for each group *type*. This is one of the determining factors behind the so-called *surface heterogeneity* of solid carbons. Given that most air-exposed solid carbons contain a nonnegligible *surface coverage* of heteroatoms, the foregoing factor also explains why the exact surface properties of a given solid carbon can only be reproduced within certain statistical limits.

Another consequence of electronegativity differences among different atoms on carbon surfaces is the fact that functional groups fall into two categories: (a) *donor* groups, which shift the electron cloud away from them, and (b) *acceptor* groups, which shift the electron cloud toward them. Donor groups contain σ- or π-electron pairs (e.g., OH, OR, O(C=O)R, NH₂, SH), which they can partially donate. Acceptor groups contain empty orbitals (e.g., (C=O)OH, (C=O)OR, (C=O)H, NO₂), which they tend to partially fill by withdrawing electrons from their substrate. Interestingly, these electron cloud shifts affect their nearest neighbors; these in turn affect the atoms behind them, and so forth, thereby leading to an "alternating polarity" rule, which helps to explain reactivity patterns in simple organic molecules [117,118]. It goes without saying that for the preceding scheme to proceed the substrate must be willing to comply with the donor–acceptor demands of the attached functional group(s). Carbon substrates generally contain aromatic portions that permit both electron cloud and charge delocalization via extended π bonding (i.e., resonance or "mesomeric" effects). The distinction between donor and acceptor groups is important in that it sets the stage for the generation of *acidity scales* for functional groups on carbon surfaces.

Acids and bases can be conveniently categorized in terms of their Brønsted or their Lewis strength. The strength of a Brønsted acid (i.e., its acidity) is not only determined by its tendency to donate a proton; there must also be an electron pair ready to accept the proton (i.e., a Brønsted base). The ease with which protons can be transferred is determined by the polarity of the participating functional groups, which (as already discussed) depends on their electron cloud donor–acceptor properties. Acceptor groups tend to build up and localize the electron cloud density around them, and this makes them less willing to release protons (i.e., weaker Brønsted acids); conversely, donor groups can spread out (delocalize) their electron pair and become more willing to release protons (i.e., stronger Brønsted acids). Carbon substrates largely determine the extent to which these effects can take place. For instance, if one compares the Brønsted acid strength of ArOH and Ar(C=O)OH (where Ar represents an aromatic substrate), one finds that the latter is more acidic because, even though it acquires charge from the substrate, it manages to spread it between its two oxygen atoms through resonance. The former is willing to give electrons but cannot spread its charge as efficiently as the latter. Nevertheless, ArOH is a stronger acid than ROH (where R represents an aliphatic substrate) because in the latter there is little possibility of electron donation or charge redistribution. The balance of all these effects is globally expressed as an acid dissociation constant, K_a. Acid dissociation constants ($K_a = [H+][Base]/[Acid]$;

$pK_a = -\log K_a$) allow a direct comparison of the relative strengths of different Brønsted acids. Brønsted bases can be simultaneously compared by using the relationship $pK_b = 14 - pK_a$ in water [117]. The value of pK_a is equivalent to the pH of the aqueous solution of the acid only in very dilute concentrations (i.e., not too far from pH = 7). At higher acid or base concentrations, the activity coefficients change and make the ratio [Base]/[Acid] $\neq$ 1. Table 2 lists approximate pK_a values (ranges) for a host of functional groups found on simple organic compounds. (Also included are pK_a values for some acids/bases used in studies of carbon surfaces.) Clearly, pK_a values are inversely proportional to the acid strength of each group. In other words, functional groups in Table 2 are listed in order of decreasing Brønsted acid strength.

It is important to note that the values listed in Table 2 represent only rough approximations to equilibrium values characteristic of each functional group. Under equilibrium conditions, acids high in Table 2 might be expected to protonate any base below them. For instance, carboxyl groups as acids ($3 \leq pK_a \leq 6$) could be expected to be able to transfer their protons to RNH_2 ($10 \leq pK_a \leq 11$) but not to Ar_2NH ($pK_a = 1$) groups. Similarly, the protonation of carboxyl groups as bases ($-7 \leq pK_a \leq -6$) with HNO_3 ($pK_a = -1.4$) should be virtually impossible; however, it could be partially effected with HCl ($pK_a = -7$) but only in nonaqueous media (because pH values beyond the range 0–14 cannot be achieved in aqueous solutions [107,124]). It is also important to realize that at equilibrium the pK_a values do not set sharp limits between the acidic and basic states of a functional group; rather, they provide estimates for the values below which the dissociated form of the acid will be the predominant species in solution. However, the dominance of a given form does not necessarily preclude the existence of the other; in fact, because of the unknown activity coefficients of the species involved [117], it is safer to assume that small amounts of the unfavored species might still be present ± 2 pK units away from those given in Table 2 [107].

The surfaces of solid carbons can contain a variety of groups whose Brønsted acid–base characteristics may or may not resemble those of simple compounds. The most common ones are illustrated in Fig. 2. As an example, oxygen functional groups could be made by oxygen atoms located *on* (e.g., carboxyl, phenol, carbonyl) or *in* (e.g., ring-bound species such as pyrans, i.e., chromenes, chromenols, chromanones and chromones ("pyrones") [125]) the edges of layers exhibiting different degrees of aromaticity. Consequently, such functional groups are likely to be subjected to severe inductive (i.e., donor–acceptor), mesomeric, steric, tautomeric (i.e., zwitterion formation by proton relocation within a heterocyclic layer) and intramolecular hydrogen-bonding effects [107,117]. Therefore, Table 2 should mainly be viewed as a guideline to establish the chemical state of possible surface functionalities existing on carbon surfaces at a given pK_a (or pH).

Acids and bases can also be classified in terms of their Lewis strength. Accordingly, any species with vacant orbitals (including protons) is viewed as a Lewis acid. The possibility that Lewis bases unrelated to heteroatoms (O, N, S) present on carbon surfaces can interact with Lewis acids (including protons) in solution [24,69,107] is examined in detail elsewhere [52] and illustrated in Section VII.

By far the most popular approach to the chemical analysis of solid carbons is that based on the titration of functional groups with bases or acids [18,36,126–128]. This approach was developed in response to an apparent failure of direct potentiometric titration curves to exhibit discernible endpoints in aqueous media [23,129,130]. (Rather than suggesting the technique's inapplicability to carbon characterization, smooth potentiometric responses were assumed to arise from a continuum of closely interacting functionalities [19,130].) The classical approach (Boehm's method) consisted of shaking (for at least 16 h) closed bottles containing unspecified amounts of carbon and of 0.05–0.1 N solutions of four bases of increasing strength [22,36]. The idea behind using bases with a wide range of pK_a values was to attempt to neutralize surface functionalities according to their acid strength, i.e., since a base of $pK_a = (pK_a)_i$ would only neutralize (i.e., form salts with) the functional groups having $pK_a \leq (pK_a)_i$. Sodium salts were preferred because (i) they do not form precipitates in the presence of gaseous CO_2 [22]; (ii) they minimize ambiguities due to the stoichiometry of surface salt formation [22]; (iii) their dissociation at the concentrations used is more complete than that of salts having cations with formal charges > +1 [113]; and (iv) alkali metal ions present the least specific interactions with carbon [131,132] and coal [133,134] surfaces. (However, more recent studies have shown that even alkali metal ions can affect the (determination of) surface properties of solid carbons [50,135].) The salts chosen were sodium bicarbonate, $NaHCO_3$; sodium carbonate, Na_2CO_3; sodium hydroxide, $NaOH$; and sodium ethoxide, $NaOC_2H_5$. The corresponding pK_a values of their conjugate acids were 6.37, 10.25, 15.74, and 20.58,

TABLE 2 Brønsted Acid–Base Nature of Possible Oxygen, Nitrogen, and Sulfur Functional Groups on Carbon Surfaces

Name of function[a]	Class[b]	Acid[c]	Base[c]	Approximate pK_a[d]
Nitro (B)	N	RNO_2H^+	RNO_2	−12
Nitro (B)	N	$ArNO_2H^+$	$ArNO_2$	−11
Hypochloric acid	A	$HClO_4$	ClO_4^-	−10
Hydroiodic acid	A	HI	I^-	−10
Cyano (B)	N	$RCNH^+$	RCN	−10
Aldehyde (B)	O	$R(C=OH^+)H$	$R(C=O)H$	−10
Ester (B)	O	$Ar(C=OH^+)OR$	$Ar(C=O)OR$	−7.4
Thiol (B)	S	RSH_2^+	RSH	−7
Carboxyl (B)	O	$Ar(C=OH^+)OH$	$Ar(C=O)OH$	−7
Aldehyde (B)	O	$Ar(C=OH^+)H$	$Ar(C=O)H$	−7
Keto (B)	O	$R(C=OH^+)R$	$R(C=O)R$	−7
Hydrochloric acid	A	HCl	Cl^-	−7
Sulfonic (A)	S	$ArSO_3H$	$ArSO_3^-$	−6.5
Ester (B)	O	$R(C=OH^+)OR$	$R(C=O)OR$	−6.5
Phenol (B)	O	$ArOH_2^+$	$ArOH$	−6.4
Carboxyl (B)	O	$R(C=OH^+)OH$	$R(C=O)OH$	−6
Keto (B)	O	$Ar(C=OH^+)R$	$Ar(C=O)R$	−6
Ether (B)	O	$Ar(O^+-H)R$	$Ar(O)R$	−6
Tertiary amine (B)	N	Ar_3NH^+	Ar_3N	−5
Aldehyde (B)	O	$H(C=OH^+)H$	$H(C=O)H$	−4
Ether (B)	O	$R(O^+-H)R$	$R(O)R$	−3.5
Hydroxyl (B)	O	$R_3COH_2^+$	R_3COH	−2
Hydroxyl (B)	O	$R_2CHOH_2^+$	R_2CHOH	−2
Hydroxyl (B)	O	$RCH_2OH_2^+$	RCH_2OH	−2
Water (B)	A	H_2OH^+ (i.e., H_3O^+)	H_2O	−1.74
Carbonamide (B)	N	$Ar(C=OH^+)NH_2$	$Ar(C=O)NH_2$	−1.5
Nitric acid	A	HNO_3	NO_3^-	−1.4
Pyrone (B)	O	$\{(=O^+-):(-OH)\}$	$\{(-O-):(=O)\}$	−1 to 2
Carbonamide (B)	N	$R(C=OH^+)NH_2$	$R(C=O)NH_2$	−0.5
Secondary amine (B)	N	Ar_2NH_2+	Ar_2NH	1
Pyrimidine (B)	N	$\{(-N=C-N-)H^+\}$	$\{(-N=C-N-)\}$	1–2
Amino (B)	N	$ArNH_3^+$	$ArNH_2$	3–5
Imino (B)	N	$ArNR_2H^+$	$ArNR_2$	3–5
Carboxyl (A)	O	$Ar(C=O)OH$	$Ar(C=O)O^-$	3–6
Pyrylium ion (A)	O	$\{-O^+=C-\}$	$\{-O-C(OH)-\}$	3–6
Formic acid	A	$H(C=O)OH$	$H(C=O)O^-$	3.77
Carboxyl (A)	O	$R(C=O)OH$	$R(C=O)O^-$	4–5

TABLE 2 (Continued)

Name of function[a]	Class[b]	Acid[c]	Base[c]	Approximate pK$_a$[d]
Alkaloid (B)	N	{(=N–)H$^+$}	{(=N–)}	5–8
Bicarbonate (B)	B	H$_2$CO$_3$	HCO$_3^-$	6.37
Thiol (A)	S	ArSH	ArS$^-$	6–8
Lactone (A)	O	{nCH$_2$–O–(C=O)}	nCH$_2$–(C=O)O$^-$	7–9
p-Nitrophenol	A	O$_2$N(C$_6$H$_4$)OH	O$_2$N(C$_6$H$_4$)O$^-$	7.15
Ammonia (B)	B	NH$_4^+$	NH$_3$	9.24
Phenol (A)	O	ArOH	ArO$^-$	8–11
Phenolic acid	A	C$_6$H$_5$OH	C$_6$H$_5$O$^-$	9.89
Nitro (A)	N	RCH$_2$NO$_2$	R(C–H$^-$)NO$_2$	10
Carbonate (B)	B	HCO$_3^-$	CO$_3^{-2}$	10.25
Tertiary amine (A)	N	R$_3$NH$^+$	R$_3$N	10–11
Amino (A)	N	RNH$_3^+$	RNH$_2$	10–11
Carbonate (B)	B	HCO$_3^-$	CO$_3^{-2}$	10.33
Hydroquinone	A	HO(C$_6$H$_4$)OH	HO(C$_6$H$_4$)O$^-$	10.35
Thiol (A)	S	RSH	RS$^-$	10–11
Oxime (A)	N	Ar=NOH	Ar=NO$^-$	10–12
Secondary amine (B)	N	R$_2$NH$_2^+$	R$_2$NH	11
Aldehyde (A)	O	Ar(C=O)H	Ar(C=O)$^-$	11–14
Ethylene glycol	A	HO(C$_2$H$_4$)OH	HO(C$_2$H$_4$)O$^-$	14.22
Hydroxide (B)	B	H$_2$O	OH$^-$	15.74
Carbanion (B)	C	Ar	Ar$^-$	16–23
Hydroxyl (A)	O	RCH$_2$OH	RCH$_2$O$^-$	16
Hydroxyl (A)	O	R$_2$CHOH	R$_2$CHO$^-$	16.5
Hydroxyl (A)	O	R$_3$COH	R$_3$CO$^-$	17
Carbonamide (B)	N	R(C=O)NH$_2$	R(C=O)NH$^-$	17
Ethoxide (B)	B	C$_2$H$_5$OH	C$_2$H$_5$O$^-$	20.58
Cyano (A)	N	RCH$_2$CN	R(C–H$^-$)CN	25
Ammonia (A)	A	NH$_3$	NH$_2^-$	38
Phenyl (A)	A	C$_6$H$_5$–H	C$_6$H$_5^-$	43
Methyl (A)	A	CH$_3$–H	CH$_3$	48

[a]As acid (A) or base (B).
[b]O, oxygen; N, nitrogen; S, sulfur. (Also included: A, acid; B, base.)
[c]R, aliphatic end; Ar, aromatic end; brackets ({ }) denote ring compounds.
[d]Experimental values (ranges) at ambient conditions, compiled from Refs. 36,107,116, 117,119–125.

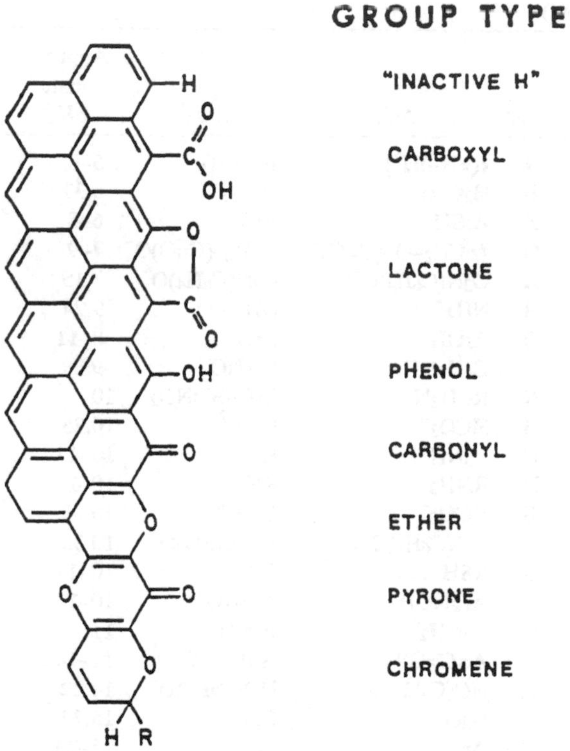

GROUP TYPE

"INACTIVE H"

CARBOXYL

LACTONE

PHENOL

CARBONYL

ETHER

PYRONE

CHROMENE

FIG. 2 Examples of oxygen and hydrogen functional groups on carbon surfaces. Active (versus inactive) hydrogen is defined [69] as that located on Brønsted acidic sites.

respectively (see Table 2), i.e., about 4.7 pK units (on average) apart from each other. Considering the pK_a values of possible functional groups on carbon surfaces (see Table 2), it was reasonable to expect (assuming that these groups behaved like those of simple organic compounds) that NaHCO$_3$ would be able to neutralize all carboxyl groups but no phenolic groups. Similarly, it was thought that the complete neutralization of phenolic groups ($8 \leq$ p$K_a \leq 11$), along with all carboxyl groups, would require an even weaker acid (or by conventional standards, a stronger base), e.g., NaOH (p$K_a = 15.74$). The possibility that groups of intermediate acid strength (i.e., between those of carboxyl and phenolic groups) existed was tested by using a base of intermediate strength, i.e., Na_2CO_3. Finally, in an attempt to account quantitatively for all the available oxygen (i.e., assuming that only oxygen functional groups were responsible for the base uptakes), $NaOC_2H_5$ was used to expand the quantification of the amount of acidic sites to include those with extremely low acidities (for which $pK_a \le 20.58$). In the latter case, experiments had to be conducted in nonaqueous media (using ethanol as solvent), and presumably under oxygen-free conditions (to avoid a base-catalyzed oxidation of the carbon [19,136]; see Section VI.A).

Initial investigations [22,36] suggested that simple stoichiometric ratios (1:2:3:4) existed among the uptakes of the increasingly strong bases used by moderately to strongly oxidized carbons. This was taken as an indication that different oxidizing agents [36] could in fact generate the same type of (surface) structure on solid carbons [22]. By combining the foregoing information with that from other analytical reactions, the following interpretation was offered:

$NaHCO_3$ titrates carboxyl groups only.
Na_2CO_3 titrates carboxyl plus lactone (i.e., ester ring) groups.
NaOH titrates carboxyl, lactone, and phenolic groups.
$NaOC_2H_5$ titrates carboxyl, lactone, phenolic, and carbonyl groups.

The content of each individual group could then be estimated by difference. Unfortunately, simple uptake ratios were not found by other workers [20,29,65,137]. Among other problems (related to chemical reactions themselves, as previously discussed), the fact remained that not all the organic oxygen could be accounted for by assuming 1:1 stoichiometric titration ratios. (The undetected balance was—and still is [13,61,65]—attributed to ether links without further chemical proof, and in a few cases to basic surface oxides [52].) Furthermore, independent investigations showed well-defined inflection points in analogous titration curves [20,138], which generally have not been observed in subsequent studies [22,30,51,115,139–141]. In addition, complications arise when trying to use as titrants either Na_2CO_3 (since it reacts slowly to form H_2CO_3) or $NaOC_2H_5$ (because its addition to carbonyl groups can be accompanied by side reactions [22,36]). An alternative characterization procedure (which we shall call Rivin's method) was later proposed by combining the information obtained from chemical and TPD experiments [19,30]. Accordingly, the distribution of specific functional group types on oxidized, untreated, and heat-treated carbons was estimated using the following assumptions [19,36]:

NaOH titrates carboxyl, lactone, and phenolic groups.
NaHCO3 titrates carboxyl groups only.
CO desorption arises from phenolic and carbonyl groups only.
CO2 desorption arises from carboxyl and lactone groups only.

The individual content of each surface group type could then be calculated as follows [12,19]:

Carboxyl groups = groups titrated with NaHCO3.
Lactone groups = groups desorbed as CO2 minus those titrated with NaHCO3.
Phenol groups = groups titrated with NaOH minus those desorbed as CO2.
Carbonyl groups = groups desorbed as CO+CO2 minus those titrated with NaOH.

The characterization of several series of pretreated carbons using Rivin's method has been addressed at length elsewhere [30]. It should be noted that these approaches are restricted to the characterization of acidic (i.e., base-consuming, presumably oxygen-containing) functional groups. Part of these groups presumably contain the so-called active (Brønsted acidic) hydrogen [69], which constitutes only a portion of the total hydrogen content of solid carbons [142]. Even though hydrogen is after oxygen, the most common heteroatom found in solid carbons, its surface coverage is generally very low. Nitrogen, sulfur, and other heteroatom-containing groups have not been investigated in as much detail, partly because they are generally minor constituents of solid carbons. (However, even minor amounts of, e.g., binary heterocyclic nitrogen groups could influence the acid–base character of a solid carbon considerably [117].) Neither have basic (acid-consuming) functionalities in general (see Table 2) received as much attention as acidic (base-consuming) groups. Both oxygen-containing and oxygen-free basic sites have been postulated to coexist on carbon surfaces; their characteristics are critically evaluated elsewhere [52]. The fact that several types of basic groups could exist on carbon surfaces invites their investigation (by analogy with Boehm's method) via titration with *acids* of increasing strength. Objections to this approach may be raised on the grounds that not enough information is available on the (organic) nature of basic groups as to permit an adequate interpretation of the results. However, one of the reasons why not enough information is available is the lack of evaluation methods for basic groups. Evaluating the acid titration ranges of otherwise well-characterized carbon surfaces would allow one to assess (i) whether distinct acidity ranges do exist and (ii) if they do, what pK_a ranges would the basic groups fall within. By comparing such information with available organic chemistry data (e.g., see Table 2), one could (in principle) develop a comprehensive scheme of possible basic functional groups analogous to the acidic group scheme produced by Boehm's method.

Table 2 includes pK_a values for some acids that have or could have been used for the characterization of carbon surfaces. For instance, HCl ($pK_a = -7$) has been used to evaluate basic groups [23,51,52,143], which, according to Table 2, could include pyrones ($-1 < pK_a < 2$), resonance-stabilized pyrylium ions ($3 < pK_a < 6$) and carbanions ($pK_a = 16–23$), and essentially every basic group with $pK_a >$ ca. -7. Similarly, formic acid ($pK_a = 3.77$) was reported to be able to titrate carbonium (e.g., pyrylium) ions [19]. In addition, there is ample evidence that part of the acid adsorption by carbons is physical in nature [18,65,131]. Early studies had attributed the increase in adsorption capacity in the order HCl $<$ HNO$_3$ $<$ HClO$_4$ (see Table 2) to (mostly) physical adsorption on weakly basic sites [144]. (The intermediate position of HNO$_3$ is consistent with its ability to react with carbon surfaces [30].) This physically adsorbed portion appears to be significantly enhanced in highly acidic (below pH = 1 to 3) solutions [19,65,143]. Below this pH range, irreversible acid adsorption/reaction seems to take place (probably in connection with an increased production of measurable concentrations of hydrogen peroxide) [65,143]. Irreversible acid adsorption/reaction appears to cause a decrease in surface activity and in subsequent (still oxygen-dependent) acid uptake, and an increase in NaOH consumption [65], due to, e.g., oxidation of chromenes to lactones [23]. Indeed, recent studies report a pH- and oxygen-dependent adsorption mechanism on two types of basic sites (present on a carbon film outgassed at 873 K), i.e., a stronger base with $pK_a = 7.4$ and a weaker base with $pK_a < 3$ [65,145]. (Both sites were titrated using increasing amounts of aqueous 0.1 N HCl.) Furthermore, the physically bound acid uptake can be reversed by washing with toluene or Lewis bases (e.g., ethers, dioxane) stronger than those found on carbon surfaces [19,23,30,52]. All these observations are consistent with the proposed existence of identifiable basic group ranges on carbon surfaces. (For instance, increasing acid amounts could be assumed to interact in stepwise fashion with chromenes–pyrylium ions [23], then pyrones [143], and then weaker Lewis bases [52].) In close analogy with Boehm's method, the identification of the strongest basic groups would require the use of nonaqueous solvents. Both toluene [19,23] and water [51,52,65,143] have been used as solvents. Water is the most common solvent employed, but it cannot be used to study the strength of acids with $pK_a < -1.74$ (see Table 2). Toluene offers several advantages over water as solvent: (i) it allows acids and bases to be studied, at the very least, in the range $-10 < pK_a < 30$ [124]; (ii) being only slightly polar, toluene competes less favorably with the acid for strong adsorption sites [18,146]; (iii) by the same token, it minimizes the physical adsorption of acids [19], thereby permitting a more direct examination of the strong adsorption sites; and (iv) it dissolves oxygen more readily than water. The latter factor is important in connection with the observation that, as mentioned previously, molecular oxygen appears to assist the consumption of acid by at least one type of basic group [30]. Incidentally, the fact that different acid adsorption models make different accounts of the simultaneous fate (if any) of oxygen also suggests that the actual process may be determined (at equilibrium) by the relative concentration of different surface basic sites [30,52].

One recurring aspect of the consumption of acids by carbons is that in many instances the process is better explained not by ion exchange or by electron pair sharing but by electron transfer mechanisms. The same applies to a number of important processes in which carbon surfaces participate as reactants or as catalysts [16,21,102,147]. These processes fall into the realm of electrochemistry (see Section VI).

Before leaving this section, a brief comment on the value of the foregoing characterization methods is in order. Despite their fundamental weaknesses, chemical methods have had a tremendous impact on every carbon scientist's perception of carbon surfaces. It is difficult to find carbon researchers from any branch of science who do not imagine carbon surfaces to be partially decorated by functional groups in a manner consistent with organic chemistry principles. From a practical standpoint, the virtue of the foregoing methods is that, regardless of their interpretation, they provide reproducible *ranges* (of acidity and/or thermal stability) within which many carbon surface properties can be expected to exhibit distinct responses. As a result, it is possible to correlate variables such as contact potential [69], contact angle [67,68,148], electrophoretic mobility [49,50], gasification rates [54], and many, many others [18,27] with the chemical and/or thermal (range of) responses of carbon surfaces (see Sections VII and VIII). Furthermore, it is in principle possible to evaluate carbon samples in order to predict their performance in applications of interest. In practice, the predictive capabilities may be limited by a number of factors. For example, solid carbons present heterogeneous distributions of surface properties, which may change with time; also, there is a degree of experimental uncertainty associated with chemical and TPD measurements. Representative examples of relationships between various measurable parameters and fundamental properties or applications of solid carbons are summarized in Sections VII and VIII and discussed in more detail elsewhere [30].

VI. CONTRIBUTION BY ELECTROCHEMISTRY TO THE STUDY OF CARBON SURFACES

The realization that carbon surfaces respond to electrochemical stimuli can be traced as far back as the early days of carbon science and the work of Faraday. Notable among the earliest attempts to rationalize the chemical behavior of carbon surfaces was the introduction of the concept that solid carbons could behave as oxygen electrodes [32,149]. This theory explained their ability to produce H_2O_2 (by an apparent reduction of molecular oxygen) in acidic solutions [32]. It also implied that carbon surfaces were capable of donating electrons, and hence of becoming positively charged [150], in solution. The reducing power of solid carbons for diverse gas- and liquid-phase applications (see Section VI.A) was also well recognized [131]. Interestingly, carbons were known to catalyze a number of oxidation reactions as well [151]. This suggested that solid carbons were also capable of accepting electrons, as expected from the existence of unpaired electrons at crystallite edges (Section IV) [112]. In spite of the succession of commercial applications that followed these discoveries [131,152], little was clarified regarding the actual role that carbon surfaces played in oxidation–reduction (redox) processes. It was mostly through chemical (see Section V) and physical (Section IV) methods that these properties of carbon surfaces were initially investigated.

Even though electrochemical methods of analysis appeared to be sufficiently well understood prior to 1960 [48], the actual application of these techniques was hampered by a series of experimental limitations [136,153]. Shortly thereafter these limitations were overcome, and new and more powerful electrochemical approaches were introduced [153]. However, by then the world's interest in carbon materials had changed. Because of the implementation of fluid catalytic cracking technologies in the 1960s, many coal scientists abandoned their efforts and initiated investigations dealing with petroleum [154]. This constituted a serious blow to carbon science, as evidenced by a decrease in the number of pertinent research publications in the mid-1960s to early 1970s. Thanks to the first Arab oil embargo, the interest in coal and carbon science was eventually rejuvenated [154,155]. Electrochemical methods were then applied more vigorously to solid carbons, but their aim was almost exclusively to improve their proven applications, e.g., as electrodes, fuel cells, and carbon fibers/composites [12,15,17,33,34,46,47,156–165]. These applications are mostly concerned with the use of highly conducting (i.e., graphitic) carbons having low surface areas and porosities. Therefore, the quantity of functional groups on their surfaces is generally low. Porous carbon powders have been given much less attention, because their use as electrodes poses both fundamental (Section VI.B) [47] and practical [30,163] problems. The following three sections will deal with how electron-transfer reactions involving carbon surfaces (Section VI.A) can be rationalized (Section VI.B) and evaluated (Section VI.C).

A. Electron-Transfer Processes

From an electrochemical perspective, solid carbons can undergo two types of reactions: surface reactions and intercalation processes [12,15,17,33,166]. In some instances these reactions occur spontaneously, and in others they must be driven by external influences (e.g., thermal or electrical energy). Surface reactions driven by electrical energy are, in general, better understood, and obey electrochemical principles reviewed in Sections VI.B and VI.C. Both thermally induced surface reactions and intercalation processes have important repercussions for the use of solid carbons and will be addressed separately in this section.

All surface reactions in this category, regardless of their driving force, differ from other chemical reactions (Section V) in that their occurrence involves a transfer of electrons (rather than ions). The transfer of electrons between participating species (*redox* process) leads to some species (the electron donors) being oxidized, while others (the electron acceptors) are simultaneously reduced. In organic chemistry, the determination of which species gets oxidized and which gets reduced is not so straightforward [107,121], but in general most oxidations involve a gain of oxygen and/or a loss of hydrogen, the converse being true for most reductions [107]. Table 3 lists some oxidation–reduction reactions reported to be catalyzed by solid carbons.

The reactions listed in Table 3, along with those in Table 2, are meant to illustrate the prodigious chemical nature of carbon surfaces. However, it must be emphasized that although these reactions could be carried out on many solid carbon surfaces, they all have specific experimental requirements that must be met before the reactions can proceed. Conditions such as pH, oxygen exposure, surface functional groups, temperature, time, and other factors can alter, or even inhibit, the reactions. In addition, redox reactions can also take place during sample transfer or storage, or they may be accompanied by irreversible chemical changes on carbon surfaces. Since the well-known ability of solid carbons to reduce impregnated (dried) metal salts [30] is paralleled by chemical changes on carbon surfaces, these reactions are excluded from Table 3. It should also be noted that, even if the experimental conditions were such as to allow the reactions listed to proceed, the reactions are likely not to go to completion and to have to compete with parallel or consecutive processes [131].

In general, it is believed that oxidation reactions are catalyzed by basic surface oxides and inhibited by acidic oxides on carbons [105]. The introduction of basic nitrogen functional groups by, e.g., NH_3 [90,105,168] or HCN [90,105] treatments also seems to promote oxidation [90,105] (and other [169]) reactions, particularly (when in liquid phase) at high solution pH values [105]. Moreover, recent reports indicate that the catalytic activity for oxidation reactions is inversely proportional to the size of the graphene layers in solid carbons [102]. As in many electrically driven redox processes (see Section VI.C), higher reaction rates are attributed to higher edge site concentrations [15,102], with individual contributions by acidic and basic sites being capable of altering the actual rates as already noted [15,102,105]. It is tempting to attribute opposite trends to reduction reactions on solid carbons, although the evidence available to substantiate such claims [15,131] is rather meager at present.

Other surface reactions of interest (particularly from the standpoint of catalyst preparation [30]) include H_2–D_2 exchange [147], ortho-para H_2 exchange [147,170], polymerization [16,131,136,147], cracking [131], isomerization [131,136], sugar inversion [131], and hydrogen peroxide production [171] and decomposition [131]. The ability of carbon

TABLE 3 Selected Oxidation/Reduction Reactions Catalyzed by Solid Carbons[a]

Oxidation in liquid phase (in the presence of oxygen)
    ferrocyanide [Fe(II)] to ferricyanide [Fe(III)] (in acidic media)
    ferrous sulfate [Fe(II)] to ferric sulfate [Fe(III)] (in acidic media)
    cobaltous chloride [Co(II)] (in NH_4OH) to hexammine cobaltic chloride
        [Co(III)]
    stannous fluoborate [Sn(II)] to stannic fluoborate [Sn(IV)]
    titannous chloride [Ti(III)] to titannic chloride [Ti(IV)]
    hydroquinone to quinone (in aqueous media)
    formic and oxalic acid to CO_2
    methanol to formaldehyde
    ethanol to acetic acid
    ethanol to (acetaldehyde + ethylene glycol + glycoxal)
    autooxidation (in alkaline media) to carbonate
    1-butylchloride to (HCl + butene + n-octane + polymer)
    sulfurous to sulfuric acid
    iodide ions to iodine
Oxidation in gas phase
    sulfur dioxide to sulfur trioxide
    hydrogen sulfide to sulfur
    halogenation of CO, SO_2 to $COCl_2$, SO_2Cl_2
    halogenation of ethyne to CHCl=CHCl and ethylene to C_2Cl_6
    hydrazine to nitrogen
    nitrogen monoxide to NO_2 (inhibited by moisture)
    (ammonia + nitrogen monoxide) to nitrogen
    dehydrogenation of hydrocarbons to unsaturated and aromatic hydro-
        carbons (at >873 K)
Reduction in liquid phase
    ferricyanide [Fe(III)] to ferrocyanide [Fe(II)] (in alkaline media)
    ferric sulfate [Fe(III)] to ferrous sulfate [Fe(II)]
    silver nitrate [Ag(I)] to silver metal [Ag(0)]
    gold cyanide [Au(I)] to gold metal [Au(0)]
    cupric chloride [Cu(II)] to cuprous chloride [Cu(I)]
    mercuric chloride [Hg(II)] to mercurous chloride [Hg(I)]
    chlorine to chloride ions (in acidic media)
    (selective) hydrogenation of olefins and ketones
Reduction in gas phase
    molecular nitrogen to cyanamide (cyanamide process)
    chlorine to hydrogen chloride
    bromine to hydrogen bromide

[a]Compiled from Refs. 13,16,19,21,23,24,31,35,36,90,102,105,110,131,147,151,152, and 167.

surfaces to catalyze reactions involving carbonium ions seems to be related to their acidity [136], although stronger evidence favors the participation of quinonoid-type complexes (as discussed later on) [16]. It has also been stated that the apparent ability of carbon surfaces to catalyze both oxidation and reduction reactions can be explained by equilibrium considerations [131]. However, it is more probable that each reaction involves a different—and sometimes irreversible—pathway [23,143].

Surface reactions resulting from thermal influences (e.g., bond breakup by heat or light) often undergo free-radical mechanisms [107]. These mechanisms differ from traditional (potential-driven) electrochemical pathways (see later sections) in that (i) they generally do not involve the participation of ions, and (ii) electrons are transferred not by themselves, but along with one or more attached atoms (i.e., as radicals). Many types of radicals are likely to be resonance-stabilized on carbon surfaces (Section IV) [39,106]. Indeed, three categories of stable surface radicals can be distinguished: cation radicals [19], neutral radicals [39], and anion radicals [110]. From the perspective of the use of solid carbons as catalyst supports [30] (see Section VII), it should be noted that surface cations and cation radicals are of importance in anion exchange processes [19,23], neutral radicals may be important for the trapping of metal carbonyl complexes, and anion radicals permit the selective substitution on carbon surfaces of virtually any desired functional group [110]. Each category is briefly discussed next.

Cation radicals have not been proposed to exist as such on solid carbon surfaces. However, the possible coexistence of radicals and holes within the same aromatic layer was inferred from IR observations [65] and from molecular orbital calculations [172]. Cations can be stabilized by resonance in the presence of acids, as oxonium ($=O^+$) or carbonium (C^+) ions [143]. (Nitronium cations—see Table 2—and cation radicals are also known in organic chemistry [107].) On carbon surfaces, it has been postulated that oxonium ions can exist as salts of *pyrylium ions* (from chromenes) [23] or *betaines* [125] (from pyrones [143]),

whereas carbonium ions coexist as resonant forms of oxonium ions (as shown previously) [23]. Both cations and cation radicals can be produced from organic compounds in electrochemical cells (Section VI.C). It was also suggested that certain types of radicals could become carbonium ions by interaction with molecular oxygen in acid solution [23], but the scheme proposed for such a process is mechanistically unlikely [125]. A more likely mechanism for such a process can be inferred from that postulated to result from exposing carbon surfaces to water vapor at mild temperatures [114]; this involves a one-electron transfer from a π orbital to a peroxy radical, i.e., $Ar + Ar\text{-}O\text{-}O^* = Ar^* + Ar^{*+}O_2^-$, i.e., a cation radical plus a superoxide ion.

The formation of pyrylium ions from chromene structures (shown previously) requires a net loss of hydride ions (H^-). It was suggested that the ability of acidified carbon surfaces to accept hydride ions even from formic acid (in the presence of oxygen; see Table 3) allowed the quantification of the number of available cations (Ar^+) [19]. The overall reaction scheme, i.e., $Ar^+A^- + HCOOH = ArH + HA + CO_2$, where A^- is an exchangeable anion, was complicated by a parallel acid-catalyzed decomposition of surface carboxyl groups. This extra production of CO_2, which had been promoted by refluxing the carbon slurries, could be accounted for by conducting the reaction in the absence of oxygen. In spite of the arbitrariness of the approach, it was shown that the semiquantitative determination of cationic sites on carbon surfaces is possible. It would be of interest to determine whether cationic sites can also be chemically stabilized in weakly acidic or even basic solutions; if so, they could provide an additional explanation for the positive net charge found on many solid carbons at pH > 7 [30,50,52,54].

Neutral radicals (unpaired electrons) are found at edges or defects on graphene-type units of solid carbons (Section IV). Because of their high reactivity, they are promptly capped by heteroatoms, in particular oxygen, upon air exposure. The ease with which unpaired (i.e., π) electrons in oxygen molecules (O=O) make them act as diradical species (*O–O*) suggests that peroxide-type functional groups could be formed as intermediates [18]; however, insufficient evidence is available to support their existence. It has also been suggested that oxygen diradicals (molecules) may interact with carbon surface radicals (Ar*) to form reversible sigma-bonded peroxyl radicals (Ar–O–O*) of *formal* charge –1 [114,173], i.e., precursors to superoxide ions (O_2^-). Indeed, several research groups have presented indirect evidence in favor of the formation of superoxide ions upon exposing carbon surfaces to molecular oxygen [23,65,90,114,145,173,174]. These species, which should be regarded as bases [65,90,121,145], would subsequently become reduced by mechanisms that may differ in aqueous and anhydrous environments [30,174]. As expected from estimates of typical radical concentrations on solid carbons (Section IV), this mode of O_2 adsorption has been reported to account for only a minor fraction (e.g., 10%) of all the O_2 adsorbed on carbons at room temperature [114]. Furthermore, most of it can be removed by evacuation or by heating in inert gas at 373 K [114,175]; however, some residual oxygen remains so strongly bound that it can only be removed by heating up to ca. 500 K [36,114].

As noted earlier (Section IV), some oxygen functional groups must be able to yield radicals (Fig. 1), which can participate in electron-transfer processes. One well-documented example is the inhibition of the thermal polymerization of styrene by surface quinonoid groups [16,21]:

Hydroquinone itself is a well-known free-radical scavenger that is added to commercial reactors when the need arises to minimize unwanted polymerization reactions [176]. (Interestingly, hydroquinone is also known to poison certain oxidation reactions over carbon surfaces, e.g., that of sulfurous to sulfuric acid [90]—see Table 3.) Similar reactions could be expected to take place under the conditions to which many carbon materials (e.g., carbon-supported catalysts [30]) are, or could be, subjected in practice.

Anion radicals are found on the surfaces of most air-exposed solid carbons (e.g., see Fig. 1). Anion radical concentrations of up to 0.1 mmol/g can be formed by heating carbons in the presence of strong electron donors (e.g., alkali metal atoms) [19,110,167]. These species can be stabilized as alkali ion salts in inert solvents [110]. Moreover, they can be subsequently reacted with a variety of reagents (containing radical-generating or electrophilic groups) so as to load the carbon surface with virtually any desired surface functional group. For instance, exposing the anion radicals to CO_2, alcohols, CO, and amines yields substituted carboxyl, phenolic, carbonyl, and amino groups (see Table 2) [110]. Substitution with organic acids, bases, quinones and hydroquinones [110], and presumably metal carbonyl complexes and mixtures of substituents is also possible. In addition, many alternative surface chemical modification methods, other than direct oxidation–reduction sequences [30], are available [177,178]. This flexibility allows the user to select the most adequate surface chemistry for a given application. For example, in the area of catalyst preparation, issues such as the affinity of one or more functional groups toward metal precursors and solvents, and the thermal and chemical promoting/inhibitory properties of the loaded substituents immediately come to mind [30,49,52] (see Section VII).

Two additional issues that have not received enough attention in the literature are (i) the nature of ion-radical groups other than quinonoid complexes on carbon surfaces, and (ii) the use of electron donors/acceptors to probe the properties of carbon surfaces. These issues deserve some comments, which are offered next.

As mentioned in Section IV, there is some evidence that groups other than quinonoid complexes (Fig. 1) might participate in electron transfer processes [23,90,100,105]. Since the nature of these complexes has not been addressed before (to the authors' knowledge), most indications of, e.g., surface oxygen radical transfer sites are generally attributed to quinonoid-type complexes only [16,21,23,35,36,100]. However, other functional groups could, in principle (and in practice [46]), intervene as well [100]. Let us start by considering the four (or five) types of oxygen functional groups titrated by Boehm's method (Section V). The manner in which carbonyl and phenolic groups could participate in direct electron transfer processes was illustrated in Fig. 1. On the other hand, lactones of the following two kinds are believed to exist on carbon surfaces [20,22,23,36]: normal (or *n*-) and fluorescein (or *f*-) lactones. Both are cyclic esters (Table 2) having six and five members per ring, respectively, and are presumably attached to aromatic substrates [23]. The six-member ring is a "benzologue" of pyrone groups (illustrated previously) [125] and could in principle behave according to

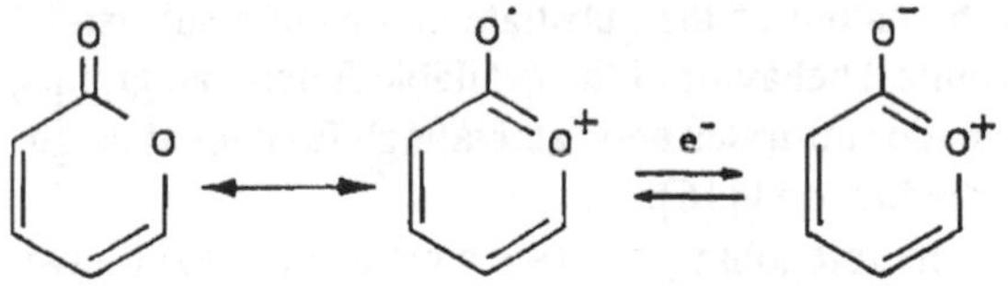

Similarly, *n*-lactones could behave according to

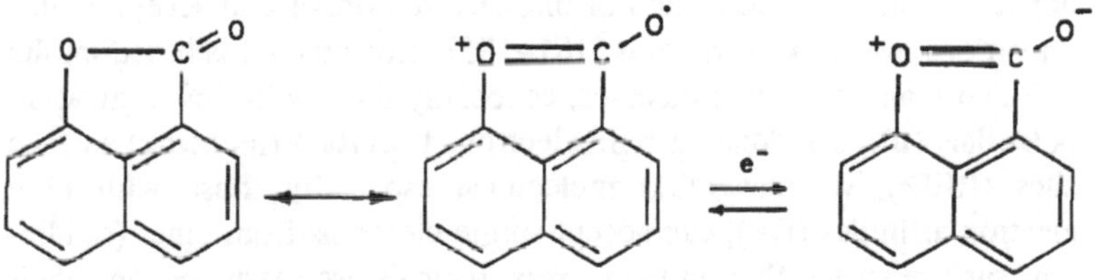

In both cases it is shown that undissociated lactones could undergo one-electron transfer processes. Since lactones remain undissociated at pH < ca. 7 (Table 2), it is expected that both processes depicted above would take place (as shown) in acidic solutions. In basic solutions, lactone rings would open [23,36] and carboxyl groups (as anions or salts, plus ketones [23,36]) would be formed. Carboxyl groups could also (in principle) resonate in concert with other functional groups, according to, e.g.,

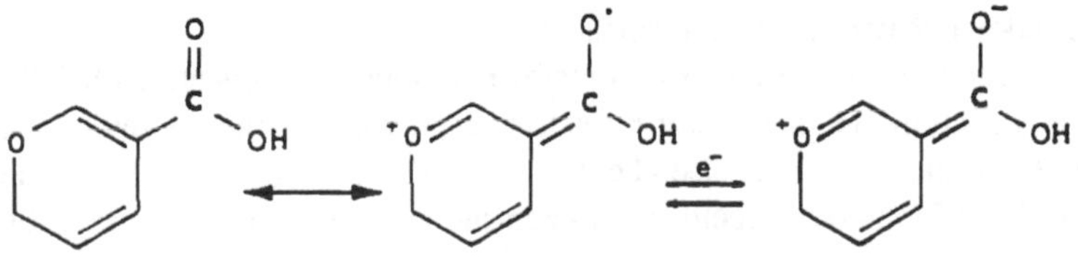

Many electrochemistry textbooks hasten to indicate that, unlike ketones, aldehydes, and other so-called electroactive groups [153,179], carboxyl groups are "inactive" electrochemically (see Section VI.C) [153]. This does not mean that carboxyl groups and the like cannot transfer electrons [180]. It simply points out that the energy needed to effect the electron transfer could actually break bonds (in simple molecules) rather than altering them as previously shown. In this regard, one must bear in mind that on solid carbons the nature of the substrate can have a substantial effect on the electrochemical behavior of the available functional groups; for instance, large aromatic substrates and "activating" functional neighbors can facilitate their reduction [116].

Attempts to quantify the foregoing processes have been based on two approaches: (i) forcing the processes to occur by applying electrical energy (as described in Section VI.C) and (ii) probing the surface with different electron donors and acceptors. The latter method has been applied to the characterization of one-electron donor and acceptor sites on various metal oxide systems [181–185]. This method is based on the fact that many organic molecules, especially those with low ionization potentials (IP), can donate single electrons to certain (reducing) surface sites (RED_i), whereas other molecules, especially those with high electron affinities (EA), can accept single electrons from other (oxidizing) surface sites (OX_i). In both cases, radicals are generated, and their concentration can be determined by ESR [182–185] or estimated using, e.g., spectrophotometry [181]. Using molecular probes of similar sizes but different IP values, both the number and the quality of the RED sites on carbon surfaces could be assessed. By the same token, probes with similar size but different EA values would allow the characterization of OX sites. In such manner, the redox properties of carbon surfaces could be evaluated just as acid–base properties are (by Boehm's method). Neither RED nor OX sites would necessarily have to be related to any of the sites described previously. However, it would not seem unreasonable to expect them to be able to coexist with [181,183–185] and to differ from [181] Brønsted sites. The foregoing method would permit an evaluation of the effects of adding different functionalities or metals on the redox behavior of carbon surfaces.

The second class of electron transfer reactions to be discussed briefly in this section are intercalation reactions. (A state-of-the-art review of this topic can be found elsewhere [186].) These reactions involve the insertion of ions or molecules between the aromatic layers of graphitic carbons [12,33,166]. Intercalation reactions rarely occur spontaneously and cannot take place readily in nongraphitic materials [12,16,21]. Intercalate layers do form on highly graphitic carbons when these are (i) chemically attacked by strong reducing or oxidizing agents (in liquid or gaseous form) or (ii) electrochemically subjected to cathodic reduction or to anodic oxidation (see Section VI.C). During cathodic reduction, the reaction $C_n + M^+ + e = M^+C_n^-$ (where M^+ is the cation of a strong electron donor—e.g., K—and e is an electron supplied by an external source) can take place. Conversely, during anodic oxidation the reaction $C_n + X^- - e = C_n^+X^-$ (where X^- is an anion stable toward oxidation – e.g., ClO_4^-) can be expected to occur. In both cases, the net charge transfer need not be complete (as suggested by the equations), and also solvent molecules can (and often do) accompany the intercalating species. As a result, intercalation compounds are quite mobile, even at room temperature [33]. However, it is their interlamellar diffusion that limits their rates of reaction, since their charge transfer processes are usually fast and reversible [12]. Naturally, the process of intercalation expands the interlayer spacing, even without altering the planarity, of the individual graphene units.

B. Electrostatic Properties

Electrostatics provides concepts borrowed from physics to account for the properties of charged particles at rest. As previously mentioned, both electron transfer (Section VI.A) and acid–base (Section V) processes can generate surface charges on solid carbons. The degree of charge localization is ultimately determined by resonance effects. For instance, carboxyl anions can distribute their negative charge between their two oxygen atoms by extended π bonding [107,121]. Hence, their charge is not "static" at a molecular level. Several functional groups on carbon surfaces are able to delocalize their charge in similar fashion. Some others, e.g., individual phenolic anions, cannot distribute their charge so readily (and hence are poorer proton donors in solution [107]). The latter kind of groups resemble more closely those found on the surfaces of ionic crystals and metal oxides. These solids, being (in general) more homogeneous than solid carbons, have also been investigated more thoroughly [126,187,188]. As a result, several theories have been introduced to explain their surface properties. Theories developed to account for the interactions of gases [189] and liquids [190] with solids were eventually applied to coals and carbons [150,191]. It was soon found that coal and carbon surfaces do not necessarily behave like those of homogeneous crystalline solids [95,132,192–194]. Perhaps the most serious limitation of older models is that they dwell on the behavior of the *fluids* instead of on the exact nature of the *surface* being characterized. Nevertheless, these models were adopted by carbon users because of their simplicity and their qualitative success in describing many surface properties of carbons (see Section VI.C) [18,146].

One popular model for the description of solid–liquid interactions is the so-called electrical double-layer model [18,113,195,196]. (Perhaps more realistic names are *triple-layer* [197] or *multilayer* [18].) Its century-old theory (developed by Helmholtz and Stern) is rooted in simple electrical principles. All solid surfaces have charge or electron imbalances due to the incomplete coordination of their outermost atoms. These imbalanced sites will thus adsorb or react with many surrounding molecules (e.g., O_2, H_2O). Their interaction lowers, but does not eliminate, the electrical differences between the surface and the bulk of the solid. In solution, the remaining surface charges will attract or repel ions and polar molecules. Additional charges (positive and negative) may develop by, e.g., surface ionization or ion adsorption. The result is that the surface and the liquid end up containing an equal number of opposite charges. As predicted by Coulomb's law, separated charges attract or repel each other with a force inversely proportional to the square of their distance. All mobile charges (ions and polar molecules in solution) perturbed by a solid would therefore tend to stick to opposite charges on its surface. An idealized version of this effect is shown in Fig. 3.

A solid surface with net negative charge is shown to attract solvated cations and polar solvent (e.g., water) molecules toward it. The latter form a solvation (e.g., hydration) sheath around the solid, which constitutes the first (or inner Helmholtz) layer on the surface. (Note that their dipoles—arrows—can point toward or away from the surface in response to local effects.) Then the solvated ions (cations in this case) approach the surface as far as the second (or outer Helmholtz) layer. Some ions (mostly anions and large cations of formal charge > +1) do not possess (complete) primary solvation sheaths [195] and can thus penetrate even deeper toward the solid. A few of these ions manage to place themselves in direct contact with the surface. They are therefore said to be *contact-adsorbed*. It is generally these ions that participate in redox processes

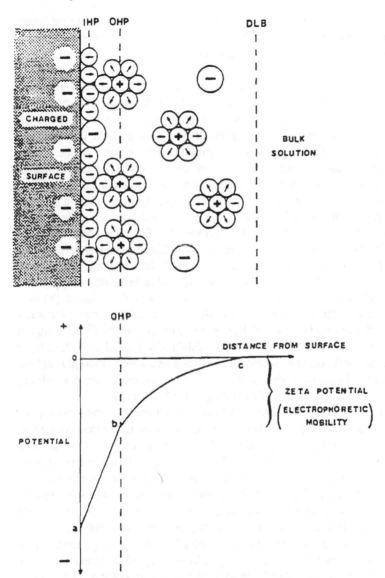

FIG. 3 Idealized structure and potential variation across an electrical double layer. A negatively charged flat surface is depicted for illustration. Polar solvent molecules (with arrows indicating dipole orientation), anions, and solvated cations are shown as spheres. The inner (IHP) and outer (OHP) Helmholtz planes and the diffuse layer boundary (DLB) are also shown.

(Table 3), whereas the others (e.g., alkali metal ions), requiring larger energies to break their solvation sheaths, remain as active spectators, or "indifferent electrolytes." Figure 3 illustrates the contact adsorption of an anion on a flat, negatively (overall) charged surface.

Because of coulombic forces of attraction, ideal surfaces would have outer Helmholtz layers filled with enough ions to balance the charges on the surface. (Such is the case with electric capacitors [113].) However, all ions and molecules have some thermal energy that induces their motion through the liquid. The thermal energy of some of the ions is large enough to make them partially overcome the surface forces of attraction. These ions then move from their electronically preferred positions to more energetically favored ones. Consequently, a concentration profile (of cations in Fig. 3) develops and decays with increasing distance from the surface. All ions (cations in Fig. 3) needed to balance the surface forces, and which are not *specifically adsorbed* between the surface and the outer Helmholtz layer, fall into an ionic atmosphere dubbed the *diffuse layer*. The electrical double layer thus comprises a "fixed" and a "diffuse" portion, whose combined thickness (Y) is a function of the ionic strength (I) of the solution. (For example, for water on a flat surface [196] at 298 K, this thickness ($Y = 0.306/I^{0.5}$) should fall between ca. 1 and 10 nm for $0.1 > I > 0.001\ M$.) Beyond the diffuse layer lies the bulk solution, where all charges move about randomly and are equally distributed within any small volume element (hence satisfying the so-called electroneutrality principle [113,195,196]).

Surface interactions of the kind already described are more often cited in terms of electric *potentials* (or work functions) rather than coulombic forces. Differences in potential ($E = \{\text{force}\} \times \{\text{distance}\}$) develop whenever charges are separated. When dealing with two or three dimensions, average potentials (arising from all surrounding charges) are invoked. In Fig. 3 only ions and polar molecules are shown as moving charges. The potentials arising from these types of interactions are very small (< 0.01 kcal/mol) [113]. Indeed, they are about 100 times lower than the energies arising from simple physical adsorption (or gas condensation) forces, i.e., 1–5 kcal/mol [113,191,198]. Much larger potentials result from electron transfer processes, e.g., 23–46 kcal/mol (i.e., "a volt or two" in electrochemical jargon [113]). These potentials are of the order of those found in chemical bonds [119].

As mentioned in Section III, the surface potentials of a variety of pretreated carbon surfaces were measured by contact potential, i.e., by measuring the work function difference between the pretreated surface and that of a reference surface (gold) [69]. In all cases, the surface potentials were typically found to be greater than that of gold by 2–11 kcal/mol. Indeed, carbon surface potentials between 4.37 and 4.76 eV (ca. 101–110 kcal/mol) were reported; interestingly, a minimum was found when the potentials were plotted as a function of the pH of the corresponding carbon slurries [69]. The positive E values for carbon relative to gold meant that some electrons could be transferred from the gold to the carbon surface. (This transfer is consistent with other reports of electron transfer from, e.g., Cu to carbon [199] and carbon to Pt [200].) The linear increase in E as the pH fell below ca. 6 was attributed to an increasing contribution by acidic surface oxides with outward-pointing dipoles. Similarly, the increase in E as the pH increased above ca. 6 was attributed to Lewis bases (C_π sites) becoming stronger, which meant that they have more delocalized electrons and a higher affinity toward protons [30,49,52]. Of course, it was necessary to assume that (i) good solid–solid contact, despite the void spaces in carbon blacks, was achieved; (ii) contact areas could be estimated from particle dimensions; and (iii) the surface charges were homogeneously distributed [69]. (No assessment of the validity of these assumptions could be made.) Considering the fact that chemically different carbon surfaces yielded potentials within a narrow range (ca. 10 kcal/mol), as they do on different crystal faces of pure metals [201], the results seemed to be quite reasonable. More importantly, these results provided new insights about the role of functional groups in determining the electrostatic properties of carbon surfaces.

One important aspect about "static" charges on solid carbons is their spatial location. Solid carbons will hold charges as functional groups on layer edges or defects, and also as (presumably) Lewis bases (or single-electron donors) capable of electron transfer in both parallel and perpendicular [27] directions with respect to the aromatic layers. Both these general site types will be distributed throughout the structure of the carbons. Except on some highly graphitic carbons (where surface functional groups could sometimes be found in patches [202]), both types of sites are often assumed to be (to a first approximation) homogeneously distributed throughout the surface. Such appears to be, in many (if not most) cases, a poor assumption [30,52]. Furthermore, the quality of otherwise equal sites located on external (e.g., geometric) and internal (i.e., intrapore) areas is likely to differ chemically [203,204].

Sites located in narrow constrictions are less likely to develop conventional double layers because of overlapping field effects. Indeed, the pK_a values of their acidic and basic groups may shift (up and down, respectively) by 1–2 pK units with respect to external surface values [204]. Model calculations suggest that charges located "deeper than a short distance" below the external surface of porous solids do not affect the properties of the external double layer [203]. However, the application of a similar model to carbon surfaces [30] suggests that double layers can be extended significantly beyond the external surface, even though their actual properties may differ somewhat inside pores. Indeed, porous electrodes make use of their extended "electrochemical area" to sustain high reaction rates (see Section VI.C) [12]. (It should be mentioned that "porous carbon electrodes" are usually made of *macroporous*, well-interconnected carbon chunks [12], instead of small, *meso-* or *microporous* particles used in adsorption processes [13,30].)

On the other hand, both double-layer models [30,203] and external surface capacitance measurements [159,162] agree in showing that the double-layer capacities (i.e., charge distributions) of solid carbons are somewhat high. In particular, the double-layer capacities of edge sites are generally much higher (20–50 $\mu F/cm^2$) than those of basal planes (3–5 $\mu F/cm^2$) [12,17,159,162]. The reasons behind these differences can be traced to two causes: (a) porosity effects and (b) surface functional groups. Porosity can lead to external charge buildup close to openings where the surface potentials overlap. Any species in these regions might feel the influence of more than one surface (but often less than one volume, as purported by fractal geometry [205]). Therefore, the coulombic forces (and potentials) generated by groups in these regions would be higher than on external surfaces [113]. However, polar functional groups attached to solid carbon surfaces should have a fixed configuration [180]. This means that they will not readily adjust to conform to an electric field. (A similar situation arises with bundles of polymers [204] and protein molecules [206], and with humic substances [207] in solution.) This fact accounts for the low dielectric constant of functionalized carbons (ca. 3) [162,208] and coals (ca. 4) [27] relative to those of similar groups in simple organic molecules (ca. 10–20) [12]. A second way in which functional groups could increase the surface capacitance is by their own oxidation–reduction (rather than simply their ionization) [162]. In this context, it has been noted that the capacitance of carbon surfaces (in particular those with edge sites) increases with increasing pH

[12,17,162]. At the same time, cyclic voltammetry peaks (see Section VI.C) that appear at low pH seem to disappear at high pH [162]. It is tempting to ascribe these effects to other functional groups (see Section VI.A) in addition to quinonoid complexes (Fig. 1). However, the previous observations are dependent on the experimental variables used for their determination (e.g., choice of electrolyte and voltage scan rate) [17,162]. The interpretation of these effects is discussed in Section VI.C.

It seems clear that electrons can be spontaneously transferred to and from carbons brought in contact with other solids. By the same token, carbons should be able to transfer electrons across a solid–liquid boundary (Fig. 3). For that to happen, three conditions must be met: (i) the solid carbon must be able to conduct electrons (and it usually is); (ii) the liquid phase must hold ions (to transport the electrons); and (iii) a high potential difference must exist, or be induced, at the solid–liquid boundary. The simple electrical double-layer model shown in Fig. 3 only accounts for low-potential static effects at solid–liquid boundaries. No account is made in the model for the structural details (e.g., porosity), the charge distribution, and the chemical nature of the surface or of its interactions with surrounding charges. Nevertheless, the electrical double-layer model has been used to explain *qualitatively* various aspects of adsorption [18] and electrokinetic (Section VI.C) phenomena. Moreover, semiquantitative approaches to establish surface charge (not individual functional group) contents have been proposed [209,210]. However, neither the double-layer model nor its intended quantification [30] go far in helping to understand how carbon surfaces *are*. Regardless, they both fare well in allowing the prediction of how carbon surfaces *behave* under different conditions. (Note the analogy with the interpretation of physical adsorption isotherms [30,95].) These predictive capabilities were proven to be very valuable for the design of mono- and bimetallic catalysts [30].

C. Electrokinetic Properties

Electrokinetic phenomena arise as a consequence of external perturbations to the electrostatic equilibrium attained between a solid and its surroundings (discussed in Section VI.B). Such electrostatic equilibrium can be disturbed in two ways: (i) by influencing the surroundings, and (ii) by influencing the solid. These aspects can be more easily visualized for solid–liquid interfaces by looking at Fig. 3. When at rest, solid–liquid electrostatic interactions give rise to an electrical double layer (EDL). If the liquid (i.e., the surroundings) and/or the solid are perturbed in any way, the EDL will change and adapt to its new situation. The liquid can be altered chemically (e.g., by adding substances to change its pH and/or its ionic strength), physically (e.g., by inducing its laminar or turbulent flow), or electrically (e.g., by imposing an electric field on the system). Similarly, the solid can be altered chemically (e.g., by fixing functional groups on its surface), physically (e.g., by polishing or roughening the surface), or electrically (e.g., by transferring electrons to or from it, i.e., by turning it into an *electrode*). Many techniques are available to allow the investigation of carbon surface properties based on such changes. However, very little effort has been made to interpret altogether the relevant information generated by these and other techniques [12,15,17,159]. Consequently, no consensus exists regarding their meaning. That is possibly the most serious limitation of electrokinetic techniques. In this section the principles, practice, and progress of some of these approaches will be examined. Three representative examples will be analyzed in detail to illustrate their usefulness for the investigation of carbon surface properties and applications [30]. These are: (i) point of zero charge (PZC) evaluations, (ii) isoelectric point (IEP) determination, and (iii) analysis by voltammetry. Even though PZC evaluations do not necessarily involve electron transfer processes, their theory is better understood by considering the properties of electrical double layers, as conceived by electrochemists (Fig. 3) [18]. The other two techniques, which are also better interpreted using Fig. 3, are driven by external energy (or *voltage*) sources. In IEP determinations, voltage is applied to the solution to induce the movement of charges (along with solids). In voltammetry, voltage is applied to the solid to induce the movement of electrons (along with ions). The proper interpretation of these techniques (and of a substantial portion of the available literature on carbon materials [12,15,17,159]) is unduly obscured by a number of factors, to be outlined and rebutted later.

To understand the meaning of PZC values, let us consider the following example. Say one adds 1 g of powdered carbon to 100 mL of pure distilled water, and the pH of the slurry decreases. The simplest explanation for this effect is that protons are given up by the carbon and go into solution. The loss of protons would make the remaining carbon surface negatively charged. Now say one adds another gram of carbon. More protons will go into solution. But since the solution already has some protons, the surface will be less willing to donate more. A third gram of carbon would donate even fewer protons to the solution. Indeed, the more acidic the solution, the harder the dissociation of acidic functional groups becomes. If more carbon is added, a point is eventually reached at which the acidic groups are virtually unable to give up more protons to the solution. Further addition of carbon does not change (significantly) the pH of the solution. If instead of carbon a simple molecule with a single functional group had been used, that pH would correspond to its pK_a. Since carbon surfaces contain a variety of acidic and basic functional groups, that pH would correspond to an *effective* pK_a. That effective pK_a (or pH) is the PZC of the carbon in question.

The EDL theory provides a more detailed definition of PZC, even without explaining what a surface site is (except that it has a charge). Figure 3 shows that negative surface charges in contact with a solution are compensated by nearby cations (or counterions). This interaction gives rise to the two layers in its EDL. (Of course, the converse situation is true for positively charged surfaces in solution.) As mentioned before, surface charges develop by ionization, ion adsorption, and/or surface dissolution processes. It is common to assume that, at least in indifferent electrolytes, ionization processes dominate the development of surface charges. (However, additional contributions by surface dissolution and/or impurities should not be ignored [188,193,211].) On carbon surfaces, Brønsted acidic (e.g., carboxyl, phenolic) groups would tend to donate their protons to water molecules, and hence become negatively charged. The excess negative surface charges would be compensated by H_3O^+ ions spread (upon reaching equilibrium) between the fixed and the diffuse portions of the EDL. The portion of protons (or other ions) remaining in the fixed layer cannot be evaluated directly [209,210]. However, their concentration in the diffuse layer is expected to decrease exponentially until reaching bulk solution values, following the Boltzmann equation (i.e., $[H^+] = [H^+]_{BULK} \exp(-FE/RT)$, where E, the potential at the solid surface, is < 0 for negatively charged surfaces [30]). In addition, the relative concentration of OH^- ions would decrease according to the relation (pH + pOH) = 14. On the other hand, neither the solvent nor the indifferent electrolytes (generally 1:1 electrolytes in low concentrations [195]) are expected to affect the properties of the double layer. (Indeed, the solvent is treated as a continuous medium that influences the EDL only through its dielectric constant [113,204].) Since protons and OH^- ions are the only species capable of being exchanged between the solid and the liquid, they are referred to as *potential-determining* ions. The reason for this denomination will become apparent upon examination of the potential distribution within the double layer (Fig. 3).

The true potential of a surface is that at the extreme of the solid itself (see point a in Fig. 3). As mentioned before, this potential cannot be measured directly; it is a difference that is (at best) measured upon contact with other charged phases. In solution, the absolute value of the potential developed between the surface (point a) and a "fixed" layer of counterions (point b) is expected to decrease linearly, as it does in a capacitor (see Fig. 3). This occurs because counterions placed in the fixed layer will balance a portion of the charge generated by the surface (thus accounting for their denomination). The difference in potential across the EDL is given by $\Delta E = (E - E_{PZC})$, where E_{PZC} is the potential of the electrode at the PZC (i.e., in the absence of a double layer, as explained later) [188]. This relation fixes the variation in potential in the EDL approximately as $\Delta E = (RT/z^{+}F)\ln\{a^{+}/a_{PZC}\}$ or $\Delta E = (RT/z^{-}F)\ln\{a_{PZC}/a^{-}\}$, depending on the activity (a = [concentration]× [activity coefficient]) and the charge (z) of the potential-determining ion [30,188,211]. (If only protons and OH^{-} ions are potential-determining, these equations lead to the well-known pH dependence of surface potential, i.e., ΔE (in V) = 0.05915(PZC-pH) [211].) It follows that $\Delta E = 0$ when $a^{+} = a_{PZC} = a^{-}$. However, ΔE can be zero only when no charge perturbation occurs at the solid–liquid interface, i.e., when there are no ions enriching the EDL relative to the bulk solution. Therefore, the PZC is the point at which the concentration of ions is such that all surface charges are effectively neutralized. Consequently, at the PZC the net surface charge of a solid is (in the absence of specific adsorption) zero. This point can be reached by adding acids or bases to carbon slurries until their surface charges are, on average, neutralized. Excess acids or bases could actually revert the overall surface charge (e.g., by activating new functional groups) [49,50]. Nonetheless, it should be clear that at the PZC carbon surfaces cannot sustain a continuous EDL.

Alternative definitions and techniques to determine PZC values are also available [53,115,209–213]. One common definition is given in terms of adsorption densities of potential–determining ions, determined by acid–base titrations [30,115,188,209–211]. The principle behind this technique is that equivalent amounts of acids and bases should be neutralized by a carbon at its PZC. Alkalimetric titrations using various concentrations of inert electrolyte allow the determination of the PZC as the point at which the net surface charge becomes independent of ionic strength [209–211]. Mass titrations [53] are an alternative to acid–base titrations and essentially consist of adding increasing amounts of powder to a liquid (usually water) until further mass additions cease to cause measurable changes in the pH of the slurry. Mass titrations can be performed in three ways: (i) direct mass titration (i.e., direct mass addition to water with or without electrolytes) [53,214]; (ii) dual mass titration (i.e., direct mass addition to solutions of preadjusted pH values, one above and one below the expected PZC) [211]; and (iii) reverse mass titration (in which the solution is gradually added to a fixed mass of carbon) [54]. Somewhat more cumbersome methods to achieve similar results include ion exchange (with PZC being the point at which no net exchange takes place) [212,213], multiple mass addition (to solutions of known pH, assuming that at the PZC little or no change in pH will occur) [214], and relative settling rates (assuming that at the PZC particles will take the longest to sink in solution) [67,68,211], among others [188,211,215–218]. Agreement among the different techniques may be expected to fall within 1–2 pH units [211,216], although discrepancies can arise because of particle size and porosity effects [68,203], hydrophobicity [67,68], specific adsorption [196,211], and, of course, impurities [188]. More studies are needed to assess properly the contribution to measured PZC values by these effects.

Other methods by which the PZC of a solid in a liquid can often be approximated are based on the application or the measurement of DC potentials [216–218]. The purpose of applying a DC potential is to set an electric field that induces the movement of charged particles relative to the liquid. Alternatively, inducing the movement of charged particles relative to the liquid sets a characteristic potential difference, which can be measured. The most common approaches can be summarized as follows [217,218]:

Particles	Fluid	Potential	Effect
Moving	Still	Applied	Electrophoresis
Moving	Still	Measured	Sedimentation potential
Still	Moving	Applied	Electroosmosis
Still	Moving	Measured	Streaming potential

Each technique is particularly suitable for specific applications [30,216–218]. Electrophoresis is preferable for the study of small (colloidal size, i.e., < 10 μm) particles. Larger (or heavier) particles can be adequately evaluated using any of the other techniques. Solid blocks (e.g., carbon pellets, fibers, and electrodes) can only be assessed using streaming potentials.

Among the foregoing techniques, electrophoresis is probably the most popular for the characterization of powdered solids [49,50,53,54, 210,217]. A simplified scheme of its experimental setup is shown in Fig. 4a. The system consists of a voltage source (V) connected to a cathode (C) and an anode (A). The electric circuit is closed by a conducting solution containing the particles to be probed. If a potential V is applied to the system, the electrodes become polarized. The anode becomes positive, and the cathode becomes negative. Anions will therefore migrate toward the anode and cations toward the cathode. If the applied potential is large enough, even suspended charged particles will migrate. The migration of charged solids is primarily controlled by three forces: (i) the electrostatic pull set by the applied potential V; (ii) the resistance to flow, which generates frictional heat; and (iii) the weight of the solid particles. Typical experimental setups require potential inputs of up to several hundred volts to overcome gravitational and viscous forces.

After the particles are set in motion, their overall charge determines their direction. (In Fig. 4a, a negatively charged particle is shown migrating toward the anode.) Furthermore, their migration velocity is related to the magnitude of their own potential. This velocity is measured by observing the motion of the particles through a calibrated microscope. Since both the electrostatic pull and the flow resistance are proportional to the cross-sectional area of the particles, all homogeneous particles of colloidal dimensions (i.e., the very light ones) would be expected to travel through the liquid at the same velocity. In practice, carbon (and other solid) particles present a distribution of velocities whose average value (v) is reported as an electrophoretic mobility (EM), i.e., a velocity normalized with respect to the applied potential. Alternatively (and perhaps more commonly), the velocity may be related to the so-called zeta potential of the particles. The zeta potential (Z) of a moving EDL is that at its "plane of shear" (as discussed later). The magnitude of Z, at an applied potential V, is related to the average velocity and the properties of the liquid medium (dielectric constant ε and viscosity μ)

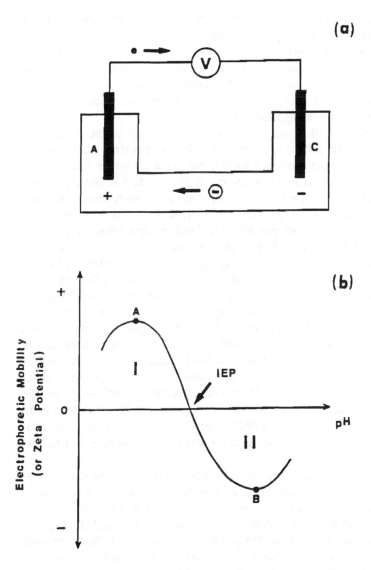

FIG. 4 Electrophoresis of solid carbons: instrument schematics (a) and typical response (b). Regions I (low pH) and II (high pH) indicate the predominance of positive (anion-adsorbing) and negataive (cation-adsorbing) surface charges.

approximately [196,218] by the expression $Z = K\mu v/(V\varepsilon)$, where K is a constant. For dilute aqueous solutions at 298 K, $Z = 199.5v/V$ (v in micrometers per second, Z and V in millivolts). Similarly, EM (in micrometers per second/volts per centimeter) = 0.078 Z (Z in millivolts). Whereas zeta potentials allow the comparison of several samples at a given applied potential, electrophoretic mobilities are needed to compare samples at different applied potentials.

Evidently, both Z and EM are indirect measures of the amount and nature of the surface charges in a solid. Amphoteric solids such as carbons have, as discussed earlier, net surface charges that strongly depend on the pH of the solution. (It is important to note, however, that *carbons are amphoteric in aqueous solution because they have separate, not coinciding, acidic and basic surface sites*, as evidenced by the pK_a values of all possible functional groups, listed in Table 2.) Hence, plots of Z or EM versus pH reveal at what conditions the surfaces in solution are positively (EM > 0) or negatively (EM < 0) charged. This effect is illustrated in Fig. 4b. The surface shown is positively charged at pH values in zone I and negatively charged at pH values in zone II. At extremely low (below point A) or extremely high (above point B) pH values, the double layers become saturated with counterions, which, by repelling each other, cause decreases in EM or Z values. (Even charge reversal might take place at extreme pH values [115], although in such cases either strong specific counterion adsorption [216,219] or even chemical attack—e.g., by the pH-adjusting additives—may have taken place.) The crossover point is the point of zero net mobility, or the *isoelectric point*. The relation between IEP and PZC is not always a direct one, as will be discussed next.

There are two important differences between the information obtained from techniques that do (e.g., electrophoresis) and do not (e.g., mass titration) require the relative movement of a phase. The first difference comes from the interpretation of the potential of a moving EDL. Figure 3 showed the EDL of a stationary solid–liquid boundary. When the solid moves (relative to the liquid), it tends to carry the fixed layer along with it. Counterions in the diffuse layer would have to reaccommodate and reach a new state of dynamic equilibrium with the moving solid. For instance, spherical suspended particles in solution would be expected to have spherical boundary layers at rest, but egg-shaped ionic atmospheres while moving [219]. As a first approximation, it can be assumed that the potential associated with the movement of the particles is that at the fixed layer (i.e., point b in Fig. 3) [196,210]. However, the potential measured by moving phase techniques will be that at an ill-defined "plane of shear" [196,211,216]. Figure 5 shows that the plane of shear (which separates the migrating from the nonmigrating portion of the EDL) may lie beyond

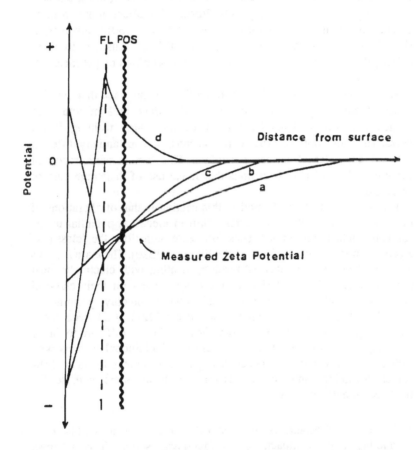

FIG. 5   Potential variation across the EDL of a moving phase. Curves: (a) E<0 (negatively charged surface), no specific adsorption; (b) E>0 (positively charged surface), excess counterion (anion) adsorption; (c) E<0 (more negatively charged than curve a), no specific adsorption; (d) E<0, excess counterion (cation) adsorption. Curves b and d exhibit surface charge reversal.

the fixed stationary layer (Fig. 3). Therefore, the relation between zeta potentials and stationary EDL potentials is difficult to predict. In fact, Fig. 5 shows three instances in which the potentials at the surface and at the fixed layer differ appreciably (curves a, b, and c), yet the measured zeta potential remains unchanged. Moreover, two curves (b and d) illustrate the effect of specific adsorption on the EDL. The specific adsorption of counterions in the fixed layer lowers its potential, to the extreme of causing a reversal of charge. Monitoring this effect allows a more careful preparation of carbon-supported catalysts [30]. However, it should be evident that even though the IEP is sensitive to these effects, the PZC might not be (especially if determined by pH measurements only).

The second difference already alluded to is the fact that moving phase techniques probe only the external surface area of immersed particles. Porous carbons have most of their area residing inside pores. Therefore, porous carbon particles can have substantially different external and internal surface potentials. This fact is very important for applications of porous carbons that make use of their high surface areas [30].

The last issue to be addressed in this section is that of alterations of the EDL driven by the direct application of electrical potential to the solids in contact with liquids. These processes constitute the "classical" electrochemical reactions found in the literature [220]. There are perhaps as many (and even as old) publications dealing with electrochemical properties of carbon materials as there are papers on their classical chemistry. Yet all their information can be understood only by a minority of carbon (and other powdered solid) users [221]. And even those who grasp the concepts from both worlds find it difficult to bring them together. One can think of a variety of additional reasons why carbon chemists and electrochemists generally appear to have difficulties in finding common grounds for discussion. Some of these difficulties are outlined next.

1. Carbon electrochemists are generally interested in highly conducting (e.g., compact, low-surface-area) solids. This narrows down their scope substantially in relation to that of other carbon users.
2. Carbon scientists have a difficult time probing the chemistry of low-surface-area materials (because of the detection limits of conventional techniques).

3. Most surface electrochemical (as well as other) techniques, when applied to heterogeneous surfaces, result in ill-defined and often irreversible reactions, and almost always involve diffusion-limited processes.
4. Most electrochemical methods are based, as far as real surfaces are concerned, on engineering-type rather than scientific approaches, and thus suffer from an inability to relate, e.g., distributed charges with localized surface functional groups.
5. Electrochemical procedures require materials with a high degree of purity, because of their high sensitivity to even minute amounts of electroactive impurities. Most commercial carbons (except perhaps for some pyrolyzed polymer blends, fibers, and carbon blacks) contain inorganic impurities, some of which could become electroactive.
6. Another common problem is one of proportions. For example, pore openings of 1 μm are considered small for electrode materials [12] but huge for most gas and liquid adsorbents.

An attempt to reconcile these issues has been made elsewhere [30]. Here, we summarize the most important facts, our findings, and our conclusions.

Solid carbons can be made to act (by suitable design [12]) as *working* (versus reference) electrodes. This means that, under particular conditions, solid carbons can conduct electrons *into* or *out of* a conducting solution (see Fig. 4). Figure 4 illustrates the behavior of a cathode (C) and an anode (A) in a closed electric circuit. If no potential E (or V in Fig. 4) is applied to the system, electrons will flow (as shown for charged particles) only if the combined potential of the redox reactions taking place in the electrodes is > 0. (It should be kept in mind that even though the flow of electrons involves 0–1 V, the movement of charged solid particles may require up to, e.g., 300 V.) If so, the reactions take place spontaneously, and there may be very small transfers of mass at each electrode. Otherwise, a potential would be required to drive the reactions. However, any species present in solution can undergo redox reactions that compete with those occurring at the surface of the electrodes. Therefore, interferences may occur if the applied potentials are high or low enough to affect redox reactions by the solvents, dissolved gases, and/or added electrolytes. This situation is illustrated next for the system quinhydrone–water–KCl. (The structure of quinhydrone was shown in Fig. 1.)

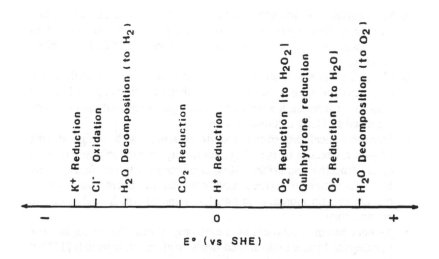

The positions assigned to each reaction, on the standard potential scale (i.e., with respect to the standard hydrogen electrode) are only relative. (Absolute values can be found elsewhere [119,123,222].) The scale shows qualitatively what might be expected upon applying a potential to the system. If the potential is shifted to the right, several reactions can take place and will do so in order of increasing potential (i.e., adsorbed oxygen might be reduced before quinhydrone groups). The solvent (water in this case) decomposes within a "window," and it is in this window that electrochemical reactions in that solvent can be studied. Indifferent electrolytes (K^+, Cl^-) fall outside the water "window" and therefore seldom interfere with simpler redox processes in solution.

When a solid carbon is immersed in an electrolyte (e.g., KCl) solution, the standard potential value ($E°$) it acquires depends on the equilibrium constant of the reactions of the potential-determining ions (e.g., protons). This value will thus depend on pH approximately as predicted by the Nernst equation, $E = E° - (0.05915/n)\text{pH}$ [113]. If no external potential is applied, $E = 0$, and $E°$ becomes a linear function of pH. Most carbons exhibit $E°$ versus pH slopes between -0.03 and -0.06 V/pH unit, suggesting that a one- or (most commonly) a two-electron transfer process occurs upon immersion. These processes are generally attributed to redox-active quinonoid [12,16,21,46] or chromene [23] groups or to mixed potentials (owing to adsorbed ion or oxygen effects) [47,127]. However, in many cases no attempts to eliminate possible complicating factors (e.g., nonequilibrium responses, peak resolution, overlapping peaks) are reported.

Of course, a transfer of electrons at carbon surfaces would involve more than simply the "jump" (or "tunneling") of electrons from one phase to the other. In the case of the quinone/hydroquinone (Q/HQ) couple, the reaction $Q + 2H^+ + 2e^- = H_2Q$ (see Fig. 1) implies that two electrons are taken up by carbonyl sites, which then combine with protons from solution to form a weak diphenolic complex of $pK_a = 10.35$ (Table 2). Such account provides a macroscopic view of the process. In reality, the kinetics [220,223] and the mechanism [108,121,222] of this reaction are not so simple [30,224]. In addition to their pH dependence, they are also sensitive to the solvent, the time of reaction, the purity of the reactants, etc. [223]. Indeed, most electron transfer processes on carbon surfaces are bound to involve the simultaneous equilibration of proton transfer processes [224]. These two types of reactions may, or may not, occur on (and/or compete for) the same site(s) on carbon surfaces. A more rigorous treatment [224,225] of the specific transfer of electrons between semiconductor surfaces and adjacent (unsolvated) ions is offered elsewhere [30]. (In this context, it is interesting to note that many nonequilibrium systems appear to be describable by equilibrium theories. An equally intriguing example is the relatively successful use of the BET or Langmuir equations in the study of gas [96] or liquid [146] adsorption on porous carbons [30].)

One point to be made with respect to indifferent electrolytes concerns the need to include them as spectators. Since electrons are carried through a liquid by ions, the more ions present the higher the conductivity of the liquid will be. The conductivity of water by itself is very low. In addition, for weakly adsorbing surfaces the concept of a surface charge is more vague than that of charges on a completely polarized system (e.g., a capacitor) [188]; however, this ambiguity is minimized by the fact that raising the ionic strength of the solution until it exceeds that of the potential-determining ions shrinks the EDL. It should be noted that the responses of charged surfaces in solution, with or without added electrolytes, will be qualitatively similar [68,219]. This observation is important for applications in which the introduction of additional substances may be undesirable (e.g., if a careful control of the chemistry of a metal precursor prior to catalyst preparation is required [30]).

After recognizing what drives electrochemical reactions and how they come about, the next question is how to monitor them. Current electrochemical techniques are concerned with the interplay among the following four variables: potential (E), current (i), concentration of reactant (C), and time (t). Examples of how these variables play a part in some techniques used for carbon characterization follow.

| Technique | Variable | | |
	Measured	Varied	Constant
Potentiometry	E	C	i (=0)
Voltammetry	i	E	C
Amperometry	i	E	t
Coulometry	i	C	E

Potentiometric analysis is equivalent to pH determinations (for which $C = [H^+]$) [109]. It is termed a steady-state electrochemical technique because no current flows through the system. All other techniques are of a transient nature. Voltammetry (which is similar to polarography) has been used more extensively for the investigation of carbon surfaces. It is reviewed in more detail elsewhere [30]. Its principles are best illustrated by an actual example reproduced in Fig. 6 [46]. Rarely do reports of more than one distinct peak (or redox couple) appear in the literature [12,17]. Nevertheless, Fig. 6 serves to illustrate several points. First, it seems possible, under appropriate experimental conditions, to detect redox processes on solid carbons. The results shown here correspond to an acid-washed, low-surface-area graphite (in 1 N H_2SO_4 solution). Peaks for higher-surface-area (porous) carbons could be expected to be more intense, and perhaps broader, assuming they contained the same redox site density as that of the graphite shown [12]. In Fig. 6, the linear voltammetric scan was started at ca. $E = 0.25$ V versus SHE (point "a"). Upon increasing the potential the current goes to zero and remains there until the reduction potential $E°$ of species in contact with, or on, the electrode is reached. At that point, current ($i_{cathodic} > 0$, by convention) begins to flow and can be recorded as a function of (i) time and (ii) potential. It takes several complete scans before equilibrium can be assumed. This is also a function of the scan rate, with faster dE/dt scans yielding larger, but often less well-resolved, peaks (as in TPD [30]). In going from point a to point b in Fig. 6, the current eventually goes up because a surface group is being reduced. In going from point b to point

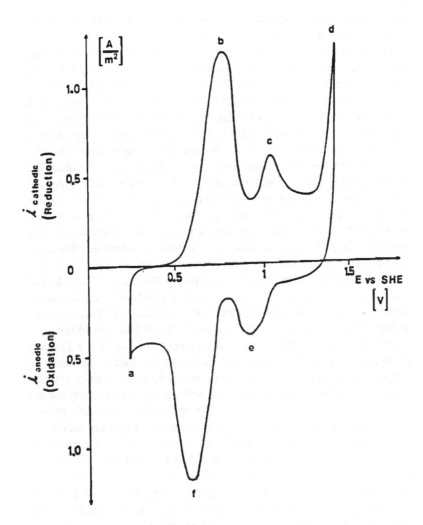

FIG. 6 Current-potential curves (cyclic voltammogram) of a graphite electrode. Adapted from Ref. 46.

c, the current initially decreases somewhat because the supply of one of the reactants (e.g., H^+) becomes scarce around the reaction site. (Note that even for nonporous graphite these processes are diffusion limited; also, stirring may improve the situation, but it will not eliminate the problem, because of its limited effect on the "fixed" part of the EDL— Fig. 3.) But then it increases again because a new reduction process is started. Indeed, as the potential is raised, a new distinct peak appears and goes through a maximum at point c. This peak decreases again because the supply of a reactant becomes rate-limiting. But upon increasing the potential, the current increases again and reaches point d, at which the decision was made to revert the sign of the potential sweep. As the negative potential is increased, the reverse (i.e., site oxidation) processes appear to take place. Peaks c and e do not differ by 30–60 mV, as they are expected to for an ideal (reversible) one- or two-electron transfer process. However, peak couples such as these are often matched, and their position is reported as a half-peak potential, i.e., $(E_c+E_e)/2$. The same applies to the redox couple made up by peaks b and f. The potential scanning cycle is ended, and recommenced, at point a. Cycling continues until equilibrium is presumed to have been reached. Typical scan rates of 2–200 mV/s can produce entire voltammograms in less than an hour.

Two clear advantages of cyclic voltammetry are that it is a fast and clean technique. It also allows one to follow the kinetic processes of simple redox couples. However, the processes in more complex systems (e.g., on rough heterogeneous surfaces with mixed types of sites) are not well understood. The situation may be alleviated if models that predict the behavior of mixed redox sites [226] are tested and proven useful.

As for the interpretation of Fig. 6 and its analogues, single peaks (or redox pairs) have been largely attributed to the reversible response of quinonoid sites (Fig. 1) in aqueous or other media [12,15,17,33,46,48]. The responses by other surface functional groups are less well known. By analogy with the behavior of functional groups in simple soluble molecules, it is often assumed that the reduction of carboxyl, lactone, phenolic, or other such group would be very difficult to attain, at least in aqueous media. This assumption may be questioned from the standpoint that surface groups in solids have electrical (if not chemical) properties that may differ considerably from those of free molecules. Besides chromene groups (Section V), only aldehydes, peroxides, various nitrogen groups (e.g., azo, azoxy, azomethine, and nitrile [153,179]), and alkenes are believed to be "electroactive" in solution. Other groups can be made electroactive by reaction with suitable reagents [153], e.g., alcohols can be oxidized to aldehydes with chromic acid, and carboxyl groups can be transformed to thiouronium salts with thiol groups identifiable by anodic voltammetry. These possibilities have not been explored for carbon surfaces. However, it is well known that carbon surface oxides are used by electrochemists as anchors for irreversible coatings of specific reactive compounds [33]. In contrast, basal planes can be coated only via reversible adsorption. Because electrode reactions are heterogeneous processes, their rates and product yields can be improved by using such chemically coated electrodes [33]. This is the electrochemically driven version of the patented functionalization approach described in Section VI.A [110].

In the absence of more firm evidence favoring other groups [23,90, 100,105], Fig. 6 was interpreted [46] by postulating that all the observed peaks are due to quinonoid-type sites being placed on different types of edges (e.g., zig-zag versus armchair) on the graphite surface. Alternatively, quinonoid sites could be located on aromatic substrates of different sizes; for instance, the E° values for molecules with carbonyl group couples attached to one (1,4-benzoquinone) and two (1,4-naphthaquinone) rings are 0.699 V and 0.480 V, i.e., over 0.2 V apart [108]. If so, one could also expect them to exhibit differences in, e.g., IR or TPD spectra [30]. However, these techniques may not be sensitive enough to detect minor variations in the properties of a limited number of functional groups. Studies with more (and selectively) oxygenated materials (even graphites) could reveal if these differences exist among quinonoid-type groups.

The final comment in this section concerns the characterization of redox sites. As mentioned earlier, it should be possible (in principle) to generate redox scales analogous to those obtained by Boehm's method. Judging from the way in which advances in other areas of chemistry were achieved, a search for such a scale would be the next logical step to be taken in this area. Many carbons exhibit electrochemical potentials of ca. 0.6–0.8 V [12], thereby suggesting that quinhydrone-type couples may be responsible for them (i.e., since various kinds of simple quinhydrones have E° values between ca. 0.5–0.7 V) [108]. However, many other carbons are reported to exhibit wider ranges of surface redox activity, falling within 0–1 V, but on occasion beyond this range [12]. What is needed is the establishment of the redox potentials of otherwise well-defined functional groups on carbon surfaces. Chromene groups, suggested as alternative redox sites [23], are not as well understood as to assign a unique redox potential to them. Yet a carbon specially prepared to contain a predominance of chromene groups could suggest in what range their potential would fall. The same applies to other functional groups. There is a need to establish experimentally whether (and under what conditions) various functional groups on carbon surfaces could become electroactive. (The current inability to achieve this objective has to do with a number of factors, among which the poor interaction between electrochemists and carbon scientists plays a key role.) Those that are electroactive should have distinguishable redox potentials associated with them. The trick is to establish these potentials. Having them, the next step would be to construct realistic speciation diagrams, i.e., "Pourbaix" diagrams [227], for carbon surfaces. In them, all participating species would be represented on a potential-pH plot, which can include oxygen and/or electron transfer dependencies. (This type of diagram has been extensively used in areas of metal corrosion and in analytical chemistry. Its principles are equally applicable to semiconductors [111].) To produce these diagrams, the first step is to list all possible reactions (both chemical and electrochemical). Then, one must find their equilibrium constants either from the literature [228] or from experiments. Even if experiments do not produce thermodynamically meaningful data, they could still provide guidelines to assess what might be expected in practice from carbon surfaces, both in chemical and electrochemical terms. On the other hand, a host of modern surface science techniques (e.g., low-energy electron diffraction and electron energy loss spectroscopy) [229] are currently being vigorously applied toward understanding the electrochemistry of so-called well-defined surfaces [230]. Thus far, out of all available solid carbons, only graphites have received their attention [229]. Perhaps in the future, technical improvements will allow the application of analogous techniques to less ordered (more common) carbon surfaces.

VII. IMPORTANCE OF CARBON SURFACE CHEMISTRY IN CATALYSIS

From our discussion in Sections III–VI, it is obvious that, despite all the advances made, the unambiguous identification of individual functional groups on carbon surfaces is a difficult task. This was also concluded, basically, at a recent workshop on interfacial phenomena at carbon surfaces, sponsored by the American Carbon Society (Atlanta, April 1992; R. T. K. Baker and D. D. Edie, organizers). Despite this state of affairs, reports of correlations between carbon surface chemistry and carbon behavior in systems of practical interest can be found in the literature. This applies both for systems in which the carbon surface interacts with inorganic gases and solutes (discussed later) and systems in which organic solutes [98,231–246] are involved. In Section VIII, we shall argue that a less detailed knowledge of carbon surface chemistry can go a long way toward explaining carbon behavior in gasification. In this section, we summarize our understanding of the behavior of carbon-supported catalysts formulated primarily with only two pieces of information, obtained in a single TPD experiment: the total oxygen coverage and the concentration of CO- and CO_2-yielding groups on the surface.

A detailed review of the use of carbon as a catalyst support is provided elsewhere [30]. Here we summarize the approach that considers carbons as charged amphoteric solids in solution. Its foundations were described in Sections III–VI, and its consequences for catalyst support applications are summarized in Table 4. The terms *acidic* and *basic* [20,36] are used to indicate negatively and positively charged surfaces, which are expected to have low (< 7) and high (> 7) IEP values, respectively [49,50]. Since even the most polar (e.g., highly oxidized) carbon surfaces contain non-polar/hydrophobic regions (as evidenced by wetting studies [67,68]), it is necessary to consider the effect of the solvent nature as well. The corresponding degree of solvent–carbon interaction in each case was determined by wetting pretreated surfaces with different solvents (or mixtures thereof) and observing their bubble evolution and sinking patterns [30]. Knowledge of the carbon type and pretreatment, the catalyst impregnation method, the solvent used, and the metal precursor allowed us to place the catalyst dispersion (fraction exposed or surface area) into one of the slots shown in Table 4. Dispersion (and thus presumably catalytic activity) was found to be relatively high or low within each box, with an apparent cutoff point at ca. 25%. The outcome of Table 4 was fully consistent with the electrostatic arguments espoused earlier [49].

In comparison with the electrostatic effects manifested upon carbon impregnation by the catalyst precursor, subsequent steps in catalyst preparation (e.g., reduction) do not necessarily have second-order effects on catalytic activity. This became clear from our studies of carbon-supported hydrodesulfurization (HDS) catalysts, in which we attempted to clarify the generally acknowledged importance of the nature of the

TABLE 4 Degree of Initial Metal Dispersion (Surface Area) Expected on Carbon Supports for Different Carbon/Solvent/Ionic Precursor Combinations

Predominant surface chemistry				Degree of solvent/C interaction	Metal dispersion	
Nature	Charge	pH$_{IEP}$	Solvent nature		Anionic precursor	Cationic precursor
Acidic	(–)	Low	P	High	Low	High
Acidic	(–)	Low	NP	Low	High	Low
Basic	(+)	High	P	Low	High	Low
Basic	(+)	High	NP	High	High	Low

P, polar; NP, nonpolar.

support [247–255] for catalytic activity. It turns out that two conflicting requirements complicate the preparation of highly active (i.e., highly dispersed) molybdenum catalysts on carbon surfaces [49,256]. On one hand, the introduction of oxygen functional groups provides anchoring sites for catalyst precursor adsorption and thus the potential for its high initial dispersion (e.g., by ion exchange). On the other hand, this introduction of mostly acidic oxygen functional groups also renders the support surface negatively charged over a wide range of pH conditions (e.g., 2–12). At very low pH, below the isoelectric point of the support, when the attractive forces prevail between the anionic Mo species (e.g., ammonium heptamolybdate) and the positively charged carbon surface, Mo polymerization seems to take place and contribute to catalyst agglomeration. In addition to this effect, the results of TPD studies, summarized in Fig. 7, provided evidence for the destruction of catalyst anchoring sites at reaction (or reduction) conditions, which leads to easier catalyst agglomeration and thus to low HDS activities. For the as-received carbon black support treated with nitric acid—to create these catalyst anchoring sites—the CO_2-yielding groups are the predominant ones, and most of them decompose at very low temperatures, in the same range as the HDS reaction temperature (400°C). It is interesting to note that the stability of the C–O surface groups can be significantly enhanced by heat treatment of the carbon black, in this case to 2500°C. Upon subsequent nitric acid treatment, a similarly large number of CO_2-yielding groups can be introduced onto the surface, but they decompose (in inert atmosphere) at temperatures that are several hundred degrees

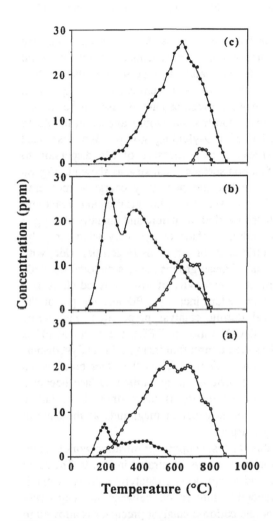

FIG. 7 Temperature-programmed desorption spectra of CO (O) and CO_2 (●) for carbon black supports (Monarch 700, Cabot): (a) as-received carbon; (b) carbon oxidized with nitric acid; (c) carbon heat-treated to 2500°C and then oxidized with nitric acid. (Reprinted with the permission of Academic Press from Ref. 49.)

higher. This approach can prove to be quite useful for some catalyst support applications [30].

Our more recent work on Mo/C systems has been directed toward the analysis not only of the quantity of oxygen functional groups that could function as anchoring sites for the catalyst precursor but also of their quality (or nature). Representative results are summarized in Fig. 8. If in fact CO_2-yielding groups are not of much use in this case—for example, because of their low thermal stability—it would appear desirable to maximize the concentration of CO-yielding groups. We first used Mo-catalyzed carbon gasification as a measure of Mo dispersion to confirm the importance of carbon surface chemistry, as shown in Fig. 8a [257]. As the carbon surface became positively charged—measured by electrophoretic mobility versus pH—the gasification reactivity increased, as expected. We then varied the nature and quantity of oxygen functional groups on the carbon surface over a very wide range by combining treatments in nitric acid, or air, or hydrogen peroxide, with the selective removal of some of these groups being monitored by TPD. Figure 8b shows the gasification reactivity of Mo-catalyzed chars as a function of the CO/CO_2 ratio determined by TPD and the pH of the impregnating solution used during catalyst preparation. For acidic carbons (that is, for low CO/CO_2 ratios in TPD), the effect of pH is relatively small; there is less than a threefold increase in catalyst dispersion as the pH is varied from 11.5 to 2.5. On the other hand, when the population of CO-yielding groups on the carbon surface becomes greater, the effect of surface chemistry is more dramatic: catalytic activity increases by more than an order of magnitude as the support surface becomes positively charged.

We have also extended this catalyst support optimization work to other systems [30]. The behavior of copper is illustrated in Fig. 9. The same general principles were found to apply, i.e., catalytic activity is maximized (Fig. 9a) when favorable interaction between the negatively charged carbon surface and the cationic catalyst precursor is allowed to occur (at high pHs). We have further complicated this system, in a realistic way, by varying the maximum catalyst reduction or heat treatment temperature (Fig. 9b). For high heat-treatment (or catalyst sintering) temperature ($\geq$923 K), the most active catalyst was the one prepared on the most acidic support. Ion exchange of Cu^{+2} cations with the large number of carboxyl groups is responsible for this effect. When the catalyst reduction temperature was low, maximum catalyst activity was

(a)

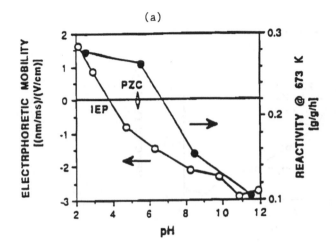

(b)

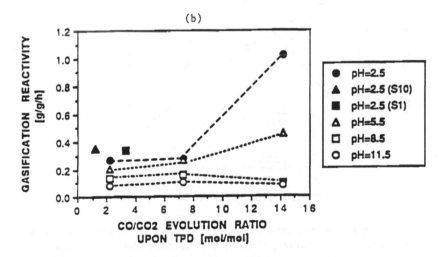

FIG. 8 (a) Correlation between surface charge and molybdenum-catalyzed carbon gasification reactivity in air (5% Mo, excess-solution impregnation). (b) Effect of carbon surface and solution chemistry on molybdenum-catalyzed gasification reactivity (and Mo dispersion). All carbons, except S10 (highly oxidized) and S1 (heat-treated), have 5% Mo and very similar surface areas and contain ~4% oxygen. From Ref. 30.

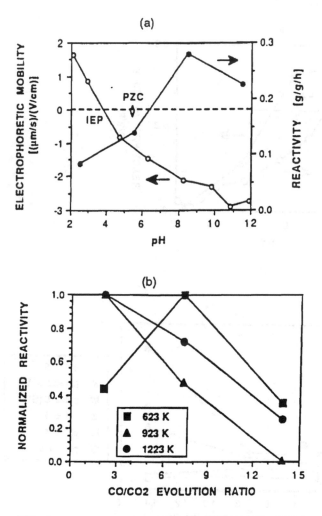

FIG. 9 (a) Correlation between surface charge and copper-catalyzed carbon gasification reactivity in air (5% Cu/Saran char, excess-solution impregnation). (b) Effect of carbon surface chemistry and heat pretreatment on copper-catalyzed gasification reactivity (and Cu dispersion). All carbons have 5% Cu and very similar surface areas and contain ~4% oxygen. From Ref. 30.

found for supports with intermediate CO/CO_2 ratios in TPD. This result is consistent with catalyst dispersion measurements (and is discussed in more detail elsewhere [30]).

Therefore, the versatility of carbon as a catalyst support [258–261] can be summarized as follows. The extent of adsorption of a catalyst precursor depends on the number of (anchoring) sites on the support. This number can be manipulated, e.g., by oxidative and/or thermal treatment of the carbon. The extent of adsorption also depends on the accessibility of these sites to the catalyst precursor. This in turn depends on the isoelectric point of the carbon, the pH of the impregnating solution, and the charge of the (ionic) catalyst precursor. The relationship between the resulting initial catalyst dispersion and the final dispersion (or catalytic activity) depends also on the thermal stability of the anchoring sites. With these variables in mind and with a given carbon support, one can *adjust the pH* during catalyst preparation to favor site accessibility and catalyst–support interaction and thus enhance catalytic activity. One can also *select judiciously the catalyst precursor* to favor these interactions. If different supports are available, one can *select or modify the carbon* to maximize the number of catalyst anchoring sites and then maximize their utility by judiciously selecting the precursor and/or adjusting the pH. This flexibility, together with the opportunity to tailor the physical properties (e.g., surface area, pore size distribution) of carbon [26,96], provides the potential to custom-tailor and fine-tune the properties of a carbon-supported catalyst to specific needs.

The positive charge on the carbon surface (see Figs. 8 and 9) is an intriguing phenomenon and deserves special attention. Organic chemists find it unlikely that protonated carboxyl groups are responsible for it (see Table 2). The literature on this topic is abundant but quite uncommitted, as discussed in Section VI. An early clue for our commitment (stated later) was some exploratory work on the palladium–carbon system [262] summarized in Table 5. This catalyst system is more flexible than the ones mentioned previously because one can work both with anionic and cationic species in solution, depending on the surface chemistry of the carbon. The results show that the use of an anionic precursor at pH<IEP produces a more pronounced rate enhancement over the uncatalyzed reaction than the use of a cationic precursor at pH>IEP. We suggested that this represented the confirmation of the importance of the following reaction for the surface of chemistry of carbons:

TABLE 5 Properties of Carbon-Supported Palladium Catalysts

Sample[a]	Pd (wt%)	I_{Pd}/I_C[b]	$t_{0.5}$[c] (min)	R[d] (h^{-1})	R (gC/gPd/h)	E_a[e] (kJ/mol)
C-HNO₃	—	—	760	0.06	—	110±4
C-HNO₃/Pd(+) [pH = 9.0]	2.0	1.53	415	0.10	5.0	119±5
C-HNO₃/Pd(+) [pH = 5.2]	3.1	1.30	600	0.08	2.6	114±8
C-HNO₃/Pd(−) [pH = 0.1]	3.0	1.68	42	0.74	24.7	140±7

[a]Pd(+), Pd(NH₃)₄(NO₃)₂; Pd(−), H₂PdCl₄; catalysts prepared by incipient wetness impregnation and reduced in hydrogen at 310°C prior to reaction.
[b]X-ray photoelectron spectroscopy intensity ratio (Pd $3d_{5/2}$)/(C 1 s).
[c]Time required to reach 50% conversion.
[d]R, Reactivity in 1 atm O₂ at 420°C (g C gasified/g C/h).
[e]Apparent activation energy (E_a) determined in the range 350–420°C for the most reactive sample and 420–450°C for the others.

$$C_\pi + 2H_2O \longleftrightarrow C_\pi-H_3O^+ + OH^-$$

We thus proposed that this reaction is responsible for providing catalyst anchoring sites on the basal planes of carbon supports, in addition to the ones associated with oxygen functional groups on the prismatic (edge) planes in the carbon structure. In more recent work, independent experimental evidence consistent with this explanation was presented [52]. This is summarized in Fig. 10. The ratio of adsorbed HCl to the oxygen content of the carbon surface is plotted as a function of the carbon's oxygen content. A ratio close to unity indicates the dominant role of pyrone-type sites in HCl adsorption, according to the following reaction:

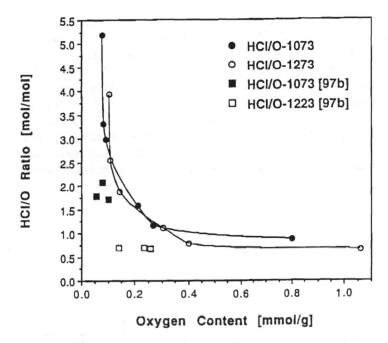

FIG. 10 Variation in HCl/O ratio with the oxygen content of carbon. Maximum TPD temperatures (used for uncovering the basic sites), in kelvins, are listed in the key. (Reprinted with the permission of Pergamon Press from Ref. 52.)

The foregoing equation cannot account for the high HCl/O ratios of carbons having low oxygen contents. Therefore, an additional HCl adsorption mechanism must be involved. After reviewing all options, it was proposed [52] that the C_π–H_3O^+ electron-donor-acceptor (EDA) complexes on the basal plane of carbon are these anchoring sites. This proposal is similar to that of Mattson and co-workers [18,232] for adsorption of phenol, in which the complexation of the –OH end of phenol with the basal plane of carbon was postulated.

VIII. IMPORTANCE OF SURFACE CHEMISTRY IN CARBON GASIFICATION

The most convenient points of departure for any level of discussion of carbon reactivity, or gasification kinetics, are two well-known and important figures, the Arrhenius plot for the carbon–oxygen reaction prepared by Smith [263], Fig. 11, and the so called "burnoff profile" obtained by Mahajan et al. [264], Fig. 12. Why do the reactivities of different carbons (having negligible levels of catalytically active impurities) vary so much at the same temperature? Most commonly, disordered (or nongraphitizable) carbons are found above the line of best fit to the data points, and more ordered (or graphitizable) carbons are found below it. Why does the reactivity of a given carbon change as a function of its conversion at constant conditions of temperature and partial pressures? Commonly, the reaction rate based on initial carbon

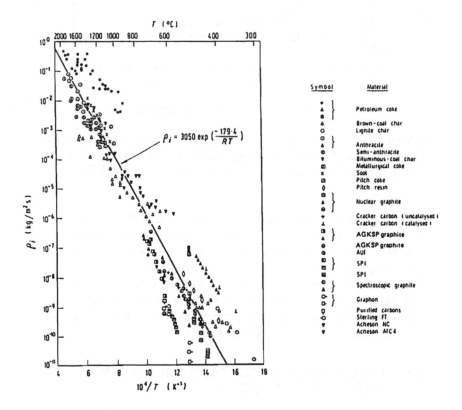

FIG. 11 Arrhenius plot of intrinsic reactivity of various carbons at 0.1 MPa oxygen. (Reprinted with the permission of Butterworths from Ref. 263.)

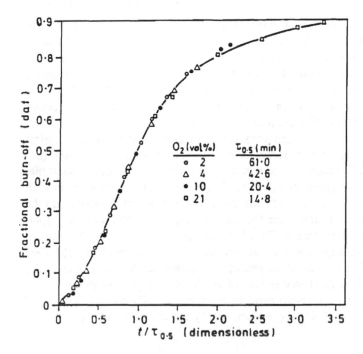

FIG. 12 Normalized reactivity plot for a coal char reacted at 405°C with oxygen (at different concentrations). (Reprinted with the permission of Butterworths from Ref. 264.)

mass (slope of Fig. 12), exhibits a maximum at intermediate conversion levels, and the rate normalized with respect to the residual carbon mass exhibits a monotonic increase with conversion.

At the most fundamental level, carbon reactivity can be expressed as follows:

[Reactivity (% h^{-1})] = [Site reactivity (h^{-1})] [Fraction of reactive sites (%)]

Are the reactivity differences simply due to changes in the number of reactive sites? Or do different carbons have different site reactivities as well? Are the reactivity changes with conversion simply due to the variations in the number of reactive sites or does the reactivity of a site depend on the degree of carbon burnoff? These are easy questions. But the answers have been quite elusive [77,81,82,101]. Nevertheless, in recent years there has been a resurgence of interest in them, and important progress toward providing some of the answers has been made. Its essence is the need [81] and the ability [82,265–269] to distinguish between at least two types of sites on the carbon surface [82,265,266]. In oxygen-transfer reactions, either stable (C–O) complexes or reactive intermediates (C(O)) are formed on these sites.

Stable carbon–oxygen complexes (or functional groups) are most conveniently titrated using temperature-programmed desorption (see Table 1). For highly reactive turbostratic carbons, it is often found that most of the oxygen on the surface forms these stable complexes; in such cases, there exists a good correlation between carbon reactivity and the integrated area under the CO and/or CO_2 desorption spectra [41,44, 82,270] as illustrated in Fig. 13. More rigorously, however, the reactive sites on the carbon surface are titrated either by deconvoluting the TPD spectra [82,269] or by means of a transient kinetics experiment [82, 265–268], as illustrated in Fig. 14 and Table 6. Figure 14 shows a very good linear relationship between reactivity in CO_2 and the fraction of reactive sites ("reactive surface area") for one coal-derived and two

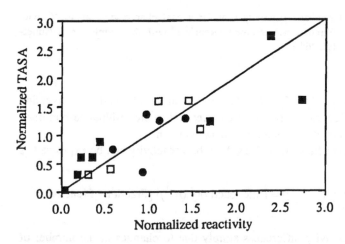

FIG. 13 Correlation of total active surface area with gasification reactivity of coal chars. □, Wang and McEnaney [271]; ●, van Heek [272]; ■, Radovic et al. [77]. (Reprinted with the permission of Kluwer Academic Publishers from Ref. 44.)

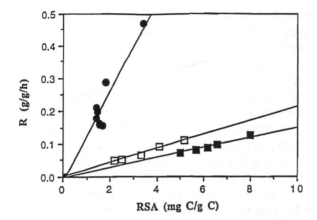

FIG. 14 Determination of turnover frequency for uncatalyzed gasification of carbon in 1 atm CO_2: ■, bituminous coal char (1093 K); □, Saran char (1133 K); ●, Carbosieve (1133 K). (Reprinted with the permission of American Chemical Society from Ref. 265.)

polymer-derived turbostratic carbons. The reactive sites (typically 1–30 out of every 1000 carbon atoms) are obviously a subset of all the surface sites (typically ~500–1000 m^2/g, i.e., 50–100 out of every 1000 carbon atoms); they can also be considered a subset of the sites that chemisorb oxygen at low temperatures (ASA, typically 10–50 out of every 1000 carbon atoms). The slopes of the straight lines shown represent the conversion-independent, temperature-dependent rate constants (site

TABLE 6 Variation of RSA-Normalized Rate Constant (TOF) with Conversion for Saran Char Gasified in 0.01 atm O_2 at 823 K

Conversion (%)	Reactivity (gC/g C/h)	RSA_{CO} (mg C/g C)	RSA_{CO_2} (mg C/g C)	RSA_{TOT} (mg C/g C)	TOF (h^{-1})
6.2	0.051	8.38	1.40	9.78	5.21
27.2	0.079	15.3	2.19	17.5	4.50
46.2	0.102	18.6	3.04	21.6	4.71
64.2	0.121	22.9	4.15	27.0	4.47
75.4	0.135	21.3	4.01	25.3	5.33

reactivities or turnover frequencies). Their differences suggest that non-catalytic carbon gasification is a structure-sensitive reaction [267], i.e., that the differences in reactivity among different carbons are not only due to variations in the number of reactive sites but also to differences in their turnover frequencies.

Similar results are shown in Table 6 for reactivity in O_2. The low pressures and high temperatures used in the experiments summarized in this table were chosen because at the more commonly used higher pressures and lower temperatures, stable (C–O) complexes occupy most of the surface and the (small number of) reactive intermediates are difficult to detect in a transient kinetics experiment.

Higher-resolution transient kinetics studies of Kapteijn and co-workers [268,273] have led to further elucidation of the mechanism of carbon gasification and quantification of the heterogeneity of reactive sites on the carbon surface. This is illustrated next.

I   II   III

From a detailed analysis of the concentration decay curves for the reactive intermediates, two types of reactive sites were postulated to exist on the carbon surface. Semiquinone structures (I and II) are thought to form on both zig-zag and armchair edge sites, whereas the less stable diketone (III) structures are thought to form at the armchair sites.

A final, less optimistic comment is warranted at the conclusion of this section. As we continue to improve our understanding of the surface chemistry of carbon in relation to its gasification reactivity [83,274], an important complication is becoming increasingly pervasive; indeed, it may make the search for some of the answers not only *e*lusive, but *ill*usive as well. We are referring to the existence of "nascent active sites" [275,276] or very reactive transient sites during carbon reactions. It is very useful to postulate their existence. We say *postulate*, because their presence cannot be directly shown. Nevertheless, an increasing number of intriguing phenomena are easily explained on this basis.

The classic example is that of coal hydropyrolysis [277]. The production of methane by the hydrogasification of carbon (coke or char) obtained after coal pyrolysis is a very slow process. Its rate can be increased dramatically if the reaction occurs not subsequent to, but during pyrolysis. Carbon atoms made available to attack by hydrogen (i.e., the nascent active sites) during these bond-breaking processes (hydropyrolysis) were found to be much more reactive than if they were given a chance to rearrange their electronic structure (e.g., by rehybridization into s^2p^2, in-plane sigma pairs [112,275,276]; see Section IV).

More recently, in our laboratory we have examined more closely [278] the enhancement of carbon reactivity toward the reduction of nitrogen oxides in the presence of oxygen [279]. The results are summarized in Table 7 and in the following figure.

The remarkable NO reduction rate enhancement in the presence of oxygen can be explained by (a) invoking the concept of nascent active sites and (b) making the necessary distinction between reactive intermediates and stable complexes on the carbon surface, as discussed previously. In the presence of O_2, the surface population of both stable complexes and reactive intermediates increases dramatically. In other words, we seem to be witnessing two superimposed effects here. First, the number of reactive sites for NO conversion increases (say by a factor of five in the absence of a copper catalyst), for example, by the following mechanism: as part of the carbon is consumed upon desorption of CO from the reactive sites, the newly created reactive sites, nascent sites, are much more susceptible to attack by NO. And second, as the oxygen coverage of the carbon surface increases (say by a factor of eight in the absence of a copper catalyst), carbon reactivity increases at least because

TABLE 7 NO_x Conversions, Carbon Conversion Rates, and Amounts of Reactive Intermediates (C(O)) and Stable Complexes (C–O) for Chars Gasified for 2 h at 300°C

		Reactant gas		
	Sample	NO (2%)	$NO(2\%)/O_2(5\%)$	O_2 (5%)
NO_x conversion (%)	Char	1	19	—
	Cu/Char	1	61	—
C conversion rate (h^{-1})	Char	0.002	0.02	0.02
	Cu/Char	0.002	0.19	0.16
Amount of C(O)	Char	0.02	0.1	0.1
(mg C/g char)	Cu/Char	0.02	0.3	0.4
Amount of C–O	Char	2	16	22
(mg C/g char)	Cu/Char	1	4	8

the number of reactive sites increases, but also because the reactivity of a site adjacent to a stable C–O complex may be higher.

This fundamental issue is one that the technique of transient kinetics is ideally suited to address and perhaps resolve.

IX. CONCLUDING REMARKS

It is difficult to comprehend how a single element can serve as a basis for so many kinds of materials. Even if one can speak of solid carbon *types* (e.g., activated carbons, carbon blacks, carbon fibers, etc.), it is useful to invoke here the analogy between *individual* solid carbons and human beings. Solid carbons—not unlike human beings—share common features (e.g., they all contain the same elements, C, H, O, N), have a similar basic structure (based on extensions of, or deviations from, a graphite skeleton), and contain some impurities ("inorganic matter"). Moreover, carbons respond to their environment in different ways, according to their origin (coal, oil, gas) and to their prior experience (age, thermal treatment, pressure).

There appear to be two levels of understanding regarding the surface properties of solid carbons: (a) fundamental and (b) applied. Each level has several sublevels, provided by contributions from an assortment of research and technical disciplines. On a fundamental level, it has been possible to establish, with a reasonable degree of confidence, a number of surface features of carbons. Arguably some of the most far-reaching advances in carbon science have been led by the evaluation of functional groups on solid carbons. This is summarized in Table 8. In the more recent past, however, these fundamental efforts have not managed to keep up with advances in carbon applications, particularly in terms of their surface chemistry (and most alarmingly in the area of electrochemistry). As a result, new and/or improved applications are much too often the result of empirical or serendipitous findings.

Before more and/or more powerful analytical techniques become available, and before predictive capabilities are developed based on both qualitative and quantitative identification of functional groups on carbon surfaces, one must resort to carefully planned experiments and judiciously chosen simplifications in order to choose the most productive courses of action regarding individual applications of carbon surfaces. Both existing and novel applications of solid carbons demand exploratory experiments aimed at comparing and contrasting results and interpretations of techniques and theories such as those discussed in this review. Such experiments should also lead to the development of a much needed, all-inclusive view of the chemistry and electrochemistry of carbon surfaces (which should illustrate, among other aspects, how all available carbon surface characterization techniques are interconnected). It is hoped that the present review will provide incentives for prospective carbon users to want to learn how to evaluate the tremendous chemical

TABLE 8 Postulated Carbon Surface Sites Associated with Brønsted, Lewis, or Redox Activity in Aqueous Solutions[a]

Group type	Donor	Indifferent	Acceptor
Brønsted proton	Phenolic, carboxyl, (lactone)	Carbonyl (ether)	(Pyrone)
Lewis electron pair	(Cπ)	(C–H)	(Cπ')
Redox electron	Quinone, (chromene)	(C–H)	Hydroquinone, (benzpyrylium ion)

[a]Groups in parentheses are those for which stronger evidence is needed to confirm their existence or their nature.

and electrochemical potential of solid carbons, in order to extend their range of applications as far as the imagination permits.

REFERENCES

1. R. H. Essenhigh, in *Chemistry of Coal Utilization*, 2nd Suppl. Vol. (M. A. Elliott, ed.), Wiley, New York, 1981, p. 1153.
2. F. J. Ceely and E. L. Daman, in *Chemistry of Coal Utilization*, 2nd Suppl. Vol. (M. A. Elliott, ed.), Wiley, New York, 1981, p. 1313.
3. J. D. Buckley, *Cer. Bull. 67*, 364 (1988).
4. D. W. McKee, in *Chemistry and Physics of Carbon*, Vol. 23 (P. A. Thrower, ed.), Marcel Dekker, New York, 1991, p. 173.
5. H. W. Kroto, J. R. Heath, S. C. O'Brien, R. F. Curl, and R. E. Smalley, *Nature 318* (6042), 162 (1985).
6. W. Krätschmer and D. R. Huffman, *Carbon 30*, 1143 (1992).
7. P. L. Walker, Jr., F. Rusinko, Jr., and L. G. Austin, *Adv. Catal. 11*, 133 (1959).
8. N. M. Laurendeau, *Prog. Energy Combust. Sci. 4*, 221 (1978).
9. F. Kapteijn and J. A. Moulijn, in *Carbon and Coal Gasification* (J. L. Figueiredo and J. A. Moulijn, eds.), Martinus Nijhoff, Dordrecht, The Netherlands, 1986, p. 291.
10. H. Marsh and K. Kuo, in *Introduction to Carbon Science* (H. Marsh, ed.), Butterworths, London, 1989, p. 107.
11. D. W. McKee and V. J. Mimeault, in *Chemistry and Physics of Carbon*, Vol. 8 (P. L. Walker, Jr., and P. A. Thrower, eds.), Marcel Dekker, New York, 1973, p. 151.
12. K. Kinoshita, *Carbon: Electrochemical and Physicochemical Properties*, Wiley, New York, 1988.
13. R. C. Bansal, J.-B. Donnet, and F. Stoeckli, *Active Carbon*, Marcel Dekker, New York, 1988.
14. H. P. Boehm, in *Structure and Reactivity of Surfaces* (C. Morterra, A. Zecchina, and G. Costa, eds.), Elsevier, Amsterdam, 1989, p. 145.
15. M. R. Tarasevich and E. I. Khrushcheva, in *Modern Aspects of Electrochemistry*, Vol. 19 (B. E. Conway, J. O'M. Bockris, and R. E. White, eds.), Plenum Press, New York, 1989, p. 295.
16. J.-B. Donnet and A. Voet, *Carbon Black: Physics, Chemistry, and Elastomer Reinforcement*, Marcel Dekker, New York, 1976.
17. J. P. Randin, in *Encyclopedia of Electrochemistry of the Elements*, Vol. VII (A. J. Bard, ed.), Marcel Dekker, New York, 1976, p. 1.

18. J. S. Mattson and H. B. Mark, Jr., *Activated Carbon: Surface Chemistry and Adsorption from Solution*, Marcel Dekker, New York, 1971.
19. (a) D. Rivin, *Rubber Chem. Technol.* 44, 307 (1971); (b) D. Rivin, *Proc. 5th Conf. Carbon*, Vol. 2, Pergamon Press, New York, 1963, p. 199.
20. B. R. Puri, in *Chemistry and Physics of Carbon*, Vol. 6 (P. L. Walker, Jr., ed.), Marcel Dekker, New York, 1970, p. 191.
21. M. L. Deviney, Jr., *Adv. Coll. Interf. Sci.* 2, 237 (1969).
22. H. P. Boehm, *Adv. Catal.* 16, 179 (1966).
23. (a) V. A. Garten and D. E. Weiss, *Rev. Pure Appl. Chem.* 7, 69 (1957); (b) V. A. Garten, D. E. Weiss, and J. B. Willis, *Aust. J. Chem.* 10, 295 (1957); (c) V. A. Garten and D. E. Weiss, *Aust. J. Chem.* 10, 309 (1957); (d) V. A. Garten and D. E. Weiss, *Aust. J. Chem.* 8, 68 (1955); (e) V. A. Garten and D. E. Weiss, *Proc. 3rd Conf. Carbon*, Pergamon Press, New York, 1959, p. 295.
24. M. L. Studebaker, *Rubber Chem. Technol.* 30, 1400 (1957).
25. I. A. S. Edwards, in *Introduction to Carbon Science* (H. Marsh, ed.), Butterworths, London, 1989, p. 1.
26. A. Linares-Solano, in *Carbon and Coal Gasification* (J. L. Figueiredo and J. A Moulijn, eds.), Martinus Nijhoff, Dordrecht, The Netherlands, 1986, p. 137.
27. D. W. van Krevelen, *Coal: Typology, Chemistry, Physics, Constitution*, Elsevier, Amsterdam, 1981.
28. H. P. Boehm and H. Knözinger, in *Catalysis: Science and Technology*, Vol. 4 (J. R. Anderson and M. Boudart, eds.), Springer-Verlag, Berlin, 1983, p. 39.
29. V. L. Snoeyink and W. J. Weber, Jr., *Prog. Surf. Membrane Sci.* 5, 63 (1972).
30. C. A. Leon y Leon D., Ph.D. Thesis, The Pennsylvania State University (1993).
31. J. W. Hassler, *Active Carbon*, Chemical Publishing Co., New York, 1951.
32. (a) A. Frumkin, *Koll. Z.* 51, 123 (1930); (b) R. Burstein and A. Frumkin, *Z. Physik. Chem.* A141, 219 (1929); (c) R. Burstein and A. Frumkin, *Dokl. Akad. Nauk. SSSR* 32, 327 (1941).
33. J. O. Besenhard and H. P. Fritz, *Angew. Chem. Internat. Edit.* 22, 950 (1983).
34. E. Yeager, J. A. Molla, and S. Gupta, *Proc. Workshop Electrochem. Carbon*, The Electrochemical Society, Pennington, NJ, 1984, p. 123.

35. (a) J.-B. Donnet, *Carbon 6*, 161 (1968); (b) J.-B. Donnet, *Carbon 20*, 267 (1982).
36. (a) H. P. Boehm, E. Diehl, W. Heck, and R. Sappok, *Angew. Chem. 76*, 742 (1964); (b) H. P. Boehm, E. Diehl, W. Heck, and R. Sappok, *Angew. Chem. Internat. Edit. 3*, 669 (1964); (c) H. P. Boehm, *Angew. Chem. Internat. Edit. 5*, 533 (1966); (d) H. P. Boehm, *Angew. Chem. 78*, 617 (1966).
37. T. C. Golden, R. G. Jenkins, Y. Otake, and A. W. Scaroni, *Proc. Workshop Electrochem. Carbon*, The Electrochemical Society, Pennington, NJ, 1984, p. 61.
38. I. L. Spain, in *Chemistry and Physics of Carbon*, Vol. 16 (P. L. Walker, Jr., and P. A. Thrower, eds.), Marcel Dekker, New York, 1981, p. 119.
39. A. Marchand, in *Carbon and Coal Gasification* (J. L. Figueiredo and J. A. Moulijn, eds.), Martinus Nijhoff, Dordrecht, The Netherlands, 1986, p. 93.
40. (a) S. Mrozowski, *Carbon 9*, 97 (1971); (b) S. Mrozowski, *Carbon 26*, 521 (1988).
41. J. D. Watt and R. E. Franklin, *Proceedings of Conference on Industrial Carbon and Graphite*, Society of Chemical Industry, London, 1958, p. 321.
42. L. Bonnetain, X. Duval and M. Letort, *Proceedings of 4th Conference on Carbon*, Pergamon Press, New York, 1960, p. 107.
43. (a) N. R. Laine, F. J. Vastola, and P. L. Walker, Jr., *J. Phys. Chem. 67*, 2030 (1963); (b) N. R. Laine, F. J. Vastola, and P. L. Walker, Jr., *Proc. 5th Conf. Carbon*, Vol. 2, Pergamon Press, New York, 1963, p. 211.
44. B. McEnaney, in *Fundamental Issues in Control of Carbon Gasification Reactivity* (J. Lahaye and P. Ehrburger, eds.), Kluwer, Dordrecht, The Netherlands, 1991, p. 175.
45. P. Ehrburger, *Carbon 29*, 763 (1991).
46. K. F. Blurton, *Electrochim. Acta 18*, 869 (1973).
47. H. Jankovska, S. Neffe, and A. Swiatkowski, *Electrochim. Acta 26*, 1861 (1981).
48. (a) J. V. Hallum and H. V. Drushel, *J. Phys. Chem. 62*, 110 (1958); (b) H. V. Drushel and J. V. Hallum, *J. Phys. Chem. 62*, 1502 (1958).
49. J. M. Solar, C. A. Leon y Leon, K. Osseo-Asare, and L. R. Radovic, *Carbon 28*, 369 (1990).

50. A. C. Lau, D. N. Furlong, T. W. Healy, and F. Grieser, *Coll. Surf. 18*, 93 (1986).
51. S. Neffe, *Carbon 25*, 441 (1987).
52. C. A. Leon y Leon, J. M. Solar, V. Calemma, and L. R. Radovic, *Carbon 30*, 797 (1992).
53. (a) J. S. Noh and J. A. Schwarz, *Carbon 28*, 675 (1990); (b) J. S. Noh and J. A. Schwarz, *Proceedings of International Carbon Conference*, Newcastle upon Tyne, UK, 1988, p. 7; (c) S. Subramanian, J. S. Noh, and J. A. Schwarz, *J. Catal. 114*, 433 (1988).
54. C. A. Leon y Leon, A. A. Lizzio, and L. R. Radovic, *Proceedings of International Carbon Conference*, Paris, France, 1990, p. 24.
55. Z.-G. Zhang, T. Kyotani, and A. Tomita, *Energy Fuels 2*, 679 (1988).
56. T. C. Brown and B. S. Haynes, *Energy Fuels 6*, 154 (1992).
57. G. Tremblay, F. J. Vastola, and P. L. Walker, Jr., *Carbon 16*, 35 (1978).
58. (a) P. J. Hall and J. M. Calo, *Energy Fuels 3*, 370 (1989); (b) P. J. Hall, J. M. Calo, and W. D. Lilly, *Proceedings of International Carbon Conference*, Newcastle upon Tyne, UK, 1988, p. 77.
59. C. A. Leon y Leon, A. W. Scaroni and L. R. Radovic, *J. Coll. Interf. Sci. 148*, 1 (1992).
60. C. Kozlowski and P. M. A. Sherwood, *Carbon 25*, 751 (1987).
61. S. R. Kelemen and H. Freund, *Energy Fuels 2*, 111 (1988).
62. B. Marchon, J. Carrazza, H. Heinemann, and G. A. Somorjai, *Carbon 26*, 507 (1988).
63. M. Starsinic, R. L. Taylor, P. L. Walker, Jr., and P. C. Painter, *Carbon 21*, 69 (1983).
64. J. J. Venter and M. A. Vannice, *Carbon 26*, 889 (1988).
65. J. Zawadzki, in *Chemistry and Physics of Carbon*, Vol. 21 (P. A. Thrower, ed.), Marcel Dekker, New York, 1989, p. 147.
66. A. S. Kotosonov, *Carbon 26*, 735 (1988).
67. (a) D. W. Fuerstenau, M. C. Williams, K. S. Narayanan, J. L. Diao, and R. H. Urbina, *Energy Fuels 2*, 237 (1988); (b) D. W. Fuerstenau, J. Diao, and J. S. Hanson, *Energy Fuels 4*, 34 (1990).
68. C. A. Leon y Leon, V. Calemma, and L. R. Radovic, *Extd. Abstracts 20th Biennial Conf. Carbon*, Santa Barbara, CA, 1991, p. 30.
69. (a) T. J. Fabish and D. E. Schleifer, *Carbon 22*, 19 (1984); (b) T. J. Fabish and D. E. Schleifer, *Proc. Workshop Electrochem. Carbon*, The Electrochemical Society, Pennington, NJ, 1984, p. 79; (c) T. J. Fabish and M. L. Hair, *J. Coll. Interf. Sci. 62*, 16 (1977).

70. O. P. Mahajan and P. L. Walker, Jr., in *Analytical Methods for Coal and Coal Products*, Vol. I (C. Karr, Jr., ed.), Academic Press, New York, 1978, p. 125.

71. F. M. Fowkes, K. L. Jones, G. Li, and T. B. Lloyd, *Energy Fuels 3*, 97 (1989).

72. M. O'Neil and J. Phillips, *J. Phys. Chem. 91*, 2867 (1987).

73. (a) A. J. Groszek, *Carbon 25*, 717 (1987); (b) A. J. Groszek, *Carbon 27*, 33 (1989).

74. M. Mermoux, Y. Chabre, and A. Rousseau, *Carbon 29*, 469 (1991).

75. (a) G. R. Hennig, in *Chemistry and Physics of Carbon*, Vol. 2 (P. L. Walker, Jr., ed.), Marcel Dekker, New York, 1966, p. 1; (b) J. M. Thomas, in *Chemistry and Physics of Carbon*, Vol. 1 (P. L. Walker, Jr., ed.), Marcel Dekker, New York, 1965, p. 121.

76. R. Phillips, F. J. Vastola, and P. L. Walker, Jr., *Carbon 8*, 197 (1970).

77. L. R. Radovic, P. L. Walker, Jr., and R. G. Jenkins, *Fuel 62*, 849 (1983).

78. S. B. Tong, P. Pareja, and M. H. Back, *Carbon 20*, 191 (1982).

79. A. Cheng and P. Harriott, *Carbon 24*, 143 (1986).

80. R. Cypres, D. Planchon, and C. Braekman-Danheux, *Fuel 64*, 1375 (1985).

81. R. L. Taylor and P. L. Walker, Jr., *Extd. Abstracts 15th Biennial Conf. Carbon*, Philadelphia, 1981, p. 437.

82. A. A. Lizzio, H. Jiang, and L. R. Radovic, *Carbon 28*, 7 (1990).

83. P. L. Walker, Jr., R. L. Taylor, and J. M. Ranish, *Carbon 29*, 411 (1991).

84. R. C. Bansal, F. J. Vastola, and P. L. Walker, Jr., *J. Coll. Interf. Sci. 32*, 187 (1970).

85. S. S. Barton, M. J. B. Evans, and B. H. Harrison, *J. Coll. Interf. Sci. 45*, 542 (1973).

86. R. H. Hurt, D. R. Dudek, J. P. Longwell, and A. F. Sarofim, *Carbon 26*, 433 (1988).

87. M. L. Studebaker and R. W. Rinehart, Sr., *Rubber Chem. Technol. 45*, 106 (1972).

88. H. Marsh, A. D. Foord, J. S. Mattson, J. M. Thomas, and E. L. Evans, *J. Coll. Interf. Sci. 49*, 368 (1974).

89. (a) P. Painter, M. Starsinic, and M. Coleman, in *Fourier Transform Infrared Spectroscopy: Applications to Chemical Systems*, Vol. 4 (J. R. Ferraro and L. J. Basile, eds.), Academic Press, New York, 1985, p. 169; (b) P. C. Painter R. W. Snyder, M. Starsinic, M. M.

Coleman, D. W. Kuehn, and A. Davis, *Appl. Spectrosc. 35*, 475 (1981).

90. B. Stöhr, H. P. Boehm, and R. Schlögl, *Carbon 29*, 707 (1991).
91. (a) E. Papirer, E. Guyon and N. Perol, *Carbon 16*, 133 (1978); (b) Hewlett-Packard Co., *ESCA Spectrometer System 5950A Operating and Service Manual*, 1972, p. 4-8; (c) P. Albers, K. Deller, B. M. Despeyroux, A. Schäfer, and K. Seibold, *J. Catal. 133*, 467 (1992).
92. T. Takahagi and A. Ishitani, *Carbon 22*, 43 (1984).
93. G. A. Somorjai, *Catal. Lett. 7*, 169 (1990).
94. (a) C. Jones and E. Sammann, *Proc. Mat. Res. Soc. Symp. 171*, 407 (1990); (b) D. L. Perry and A. Grint, *ACS Preprints, Div. Fuel Chem. 31* (1), 107 (1986).
95. H. Marsh, *Carbon 25*, 49 (1987).
96. F. Rodriguez-Reinoso, in *Carbon and Coal Gasification* (J. L. Figueiredo and J. A. Moulijn, eds.), Martinus Nijhof, Dordrecht, The Netherlands, 1986, p. 601.
97. (a) E. Papirer, S. Li, and J. B. Donnet, *Carbon 25*, 243 (1987); (b) E. Papirer, J. Dentzer, S. Li, and J. B. Donnet, *Carbon 29*, 69 (1991).
98. B. R. Puri, *ACS Symp. Ser. 8*, 212 (1975).
99. (a) T. Morimoto and K. Miura, *Langmuir 1*, 658 (1985); (b) T. Morimoto and K. Miura, *Langmuir 2*, 43 (1986); (c) K. Miura and T. Morimoto, *Langmuir 2*, 824 (1986); (d) K. Miura and T. Morimoto, *Langmuir 4*, 1283 (1988).
100. (a) J. R. Harbour and M. J. Walzak, *Carbon 22*, 191 (1984); (b) J. R. Harbour and M. J. Walzak, *Carbon 23*, 687 (1985); (c) J. R. Harbour and M. J. Walzak, *Carbon 24*, 743 (1986); (d) J. R. Harbour, M. J. Walzak, and P. Julien, *Carbon 23*, 185 (1985); (e) J. R. Harbour, M. J. Walzak, W. Limburg, and J. Yanus, *Carbon 24*, 725 (1986).
101. R. L. Taylor, Ph.D. Thesis, The Pennsylvania State University (1982).
102. H. Jüntgen and H. Kühl, in *Chemistry and Physics of Carbon*, Vol. 22 (P. A. Thrower, ed.), Marcel Dekker, New York, 1989, p. 145.
103. O. C. Cariaso and P. L. Walker, Jr., *Carbon 13*, 233 (1975).
104. J. A. Halstead, R. Armstrong, B. Pohlman, S. Sibley, and R. Maier, *J. Phys. Chem. 94*, 3261 (1990).

105. H. P. Boehm, G. Mair, T. Stoehr, A. R. de Rincon, and B. Tereczki, *Fuel 63*, 1061 (1984).

106. (a) A. R. Forrester, J. M. Hay, and R. H. Thomson, *Organic Chemistry of Stable Free Radicals*, Academic Press, New York, 1968; (b) S. E. Stein and D. M. Golden, *J. Org. Chem. 42*, 839 (1977).

107. J. March, *Advanced Organic Chemistry*, 3rd Ed., Wiley, New York, 1985.

108. G. J. Janz and D. J. G. Ives, in *Reference Electrodes: Theory and Practice* (D. J. G. Ives and G. J. Janz, eds.), Academic Press, New York, 1961, p. 270.

109. D. Midgley and K. Torrance, *Potentiometric Water Analysis*, 2nd Ed., Wiley, New York, 1991.

110. (a) D. Rivin and J. Aron, U.S. Patent 3,479,300 (1969); (b) D. Rivin and J. Aron, U.S. Patent 3,479,299 (1969).

111. S. M. Park and M. E. Barber, *J. Electroanal. Chem. 99*, 67 (1979).

112. R. V. Culver and H. Watts, *Rev. Pure Appl. Chem. 10*, 95 (1960).

113. I. N. Levine, *Physical Chemistry*, 3rd Ed., McGraw-Hill, New York, 1988.

114. C. R. Nelson, in *Chemistry of Coal Weathering* (C. R. Nelson, ed.), Elsevier, Amsterdam, 1989, p. 1.

115. (a) A. S. Arico, V. Antonucci, L. Pino, P. L. Antonucci, and N. Giordano, *Carbon 28*, 599 (1990); (b) A. S. Arico, V. Antonucci, M. Minutoli, and N. Giordano, *Carbon 27*, 337 (1989).

116. N. D. Cheronis and T. S. Ma, *Organic Functional Group Analysis by Micro and Semimicro Methods*, Wiley Interscience, New York, 1964.

117. D. D. Perrin, B. Dempsey, and E. P. Serjeant, *pK_a Prediction for Organic Acids and Bases*, Chapman and Hall, London, 1981.

118. T.-L. Ho, *Polarity Control for Synthesis*, Wiley, New York, 1991.

119. *CRC Handbook of Chemistry and Physics*, 71st Ed. (D. R. Lide, ed.), CRC Press, Boca Raton, FL, 1990–1991.

120. *Perry's Chemical Engineering Handbook*, 6th Ed. (R. H. Perry, D. W. Green, and J. O. Maloney, eds.), McGraw-Hill, New York, 1984.

121. H. Meislich, H. Nechamkin, and J. Sharefkin, *Organic Chemistry*, Schaum's Outline Series, McGraw-Hill, New York, 1977.

122. S. J. Weininger, *Contemporary Organic Chemistry*, Holt, Rinehart and Winston, New York, 1972.

123. (a) J. A. Dean (ed.), *Lange's Handbook of Chemistry*, 13th Ed., McGraw-Hill, New York, 1985; (b) E. P. Serjeant and B. Dempsey, *Ionization Constants of Organic Acids in Aqueous Solution*, IUPAC Chemical Data Series 23, Pergamon Press, New York, 1979; (c) E. Sondheimer, *J. Amer. Chem. Soc. 75*, 1507 (1953); (d) A. Williams, *J. Amer. Chem. Soc. 93*, 2733 (1971).

124. R. P. Bell, *Acids and Bases: Their Quantitative Behaviour*, 2nd Ed., Methuen, London, 1969.

125. (a) P. J. Brodgen, C. D. Gabbut, and J. D. Hepworth, in *Comprehensive Heterocyclic Chemistry*, Vol. 3 (A. J. Boulton and A. McKillop, eds.), Pergamon Press, Oxford, 1984, p. 573; (b) G. P. Ellis, in *Comprehensive Heterocyclic Chemistry*, Vol. 3 (A. J. Boulton and A. McKillop, eds.), Pergamon Press, Oxford, 1984, p. 647; (c) G. P. Ellis (ed.), *Chromenes, Chromanones and Chromones*, Wiley Interscience, New York, 1977; (d) S. Wawzonek, in *Heterocyclic Compounds*, Vol. 2 (R. C. Elderfield, ed.), Wiley, New York, 1951, p. 277; (e) J. Fried, *Heterocyclic Compounds*, Vol. 1 (R. C. Elderfield, ed.), Wiley, New York, 1950, p. 343.

126. K. Tanabe, *Solid Acids and Bases—Their Catalytic Properties*, Academic Press, New York, 1970.

127. J. van Driel, in *Fundamentals of Adsorption* (A. I. Liapis, ed.), Proceedings of 2nd Engineering Foundation Conference, United Engineering Trustees, New York, 1987, p. 589.

128. J. van Driel, in *Activated Carbon—A Fascinating Material* (A. Capelle and F. de Vooys, eds.), Norit, Amersfoort, 1983, p. 40.

129. (a) D. S. Villars, *J. Amer. Chem. Soc. 69*, 214 (1947); (b) D. S. Villars, *J. Amer. Chem. Soc. 70*, 3655 (1948).

130. M. L. Studebaker, *Proceedings of 5th Conference on Carbon*, Vol. 2, Pergamon Press, New York, 1963, p. 189.

131. J. W. Hassler, *Activated Carbon*, Chemical Publishing Co., New York, 1963.

132. (a) W. J. Weber, Jr., in *Principles and Applications of Water Chemistry* (S. D. Faust and J. V. Hunter, eds.), Wiley, New York, 1967, p. 89; (b) M. Bier and F. C. Cooper, in *Principles and Applications of Water Chemistry* (S. D. Faust and J. V. Hunter, eds.), Wiley, New York, 1967, p. 217.

133. (a) J. A. L. Campbell and S. C. Sun, *The Electrokinetic Behavior of Anthracite Coals and Lithotypes*, Special Research Report SR-44,

The Pennsylvania State University, University Park, PA, 1964;
(b) J. A. L. Campbell and S. C. Sun, *An Electrokinetic Study of
Bituminous Coal Froth Flotation and Flocculation*, Special
Research Report SR-74, The Pennsylvania State University, University Park, PA, 1969.

134. K. B. Quast and D. J. Readett, *Adv. Coll. Interf. Sci. 27*, 169 (1987).

135. S. S. Barton, B. H. Harrison, L. Pink, J. R. Sellors, and D. Kerur, *Extd. Abstracts 14th Biennial Conference on Carbon*, University Park, PA, 1979, p. 1.

136. (a) P. H. Given and L. W. Hill, *Carbon 7*, 649 (1969); (b) P. H. Given and L. W. Hill, *Carbon 6*, 525 (1968).

137. (a) V. L. Snoeyink and W. J. Weber, Jr., *Env. Sci. Technol. 1*, 228 (1967); (b) V. L. Snoeyink and W. J. Weber, Jr., in *Adsorption from Aqueous Solution*, Adv. Chem. Ser. No. 79, American Chemical Society, Washington, DC, 1968, p. 112.

138. B. R. Puri, D. D. Singh, and L. R. Sharma, *J. Ind. Chem. Soc. 34*, 357 (1957).

139. E. Papirer and E. Guyon, *Carbon 16*, 127 (1978).

140. E. Taranenko and Y. Chikhman, *Proceedings of International Carbon Conference*, Paris, France, 1990, p. 138.

141. P. Vinke, Ph.D. Thesis, Technische Universiteit Delft, The Netherlands (1991).

142. E. Papirer, J.-B. Donnet, and J. Heinkele, *J. Chim. Phys. 68*, 581 (1971).

143. (a) H. P. Boehm and M. Voll, *Carbon 8*, 227 (1970); (b) M. Voll and H. P. Boehm, *Carbon 8*, 741 (1970); (c) M. Voll and H. P. Boehm, *Carbon 9*, 473 (1971); (d) M. Voll and H. P. Boehm, *Carbon 9*, 481 (1971).

144. B. Steenberg, *Adsorption and Exchange of Ions on Activated Charcoal*, Almquist & Wiksells, Uppsala, 1944.

145. J. Zawadzki and S. Biniak, *Pol. J. Chem. 62*, 195 (1988).

146. R. L. Bond (ed.), *Porous Carbon Solids*, Academic Press, London, 1967.

147. A. J. Bird, in *Catalyst Supports and Supported Catalysts: Theoretical and Applied Concepts* (A. B. Stiles, ed.), Butterworths, Boston, 1987, p. 107.

148. K. J. Hüttinger, S. Höhmann-Wein, and G. Krekel, *Carbon 29*, 1281 (1991).

149. A. Kellerman and E. Lange, *Koll. Z. 90*, 89 (1940).
150. (a) I. Ogawa, *Biochem Z. 172*, 249 (1926); (b) H. R. Kruyt and G. S. de Kadt, *Koll. Z. 47*, 44 (1929).
151. A. B. Lamb and L. W. Elder, Jr., *J. Amer. Chem. Soc. 53*, 137 (1931).
152. V. R. Deitz, *Bibliography of Solid Adsorbents*, National Bureau of Standards, Washington, DC, 1944.
153. H. H. Willard, L. L. Merritt, Jr., J. A. Dean, and F. A. Settle, Jr., *Instrumental Methods of Analysis*, 7th Ed., Wadsworth Publishing Co., Belmont, CA, 1988, p. 697.
154. I. Wender, L. A. Heredy, M. B. Neuworth, and I. G. C. Dryden, in *Chemistry of Coal Utilization* (M. A. Elliott, ed.), Wiley, New York, 1981 p. 425.
155. H. H. Schobert, *Coal: The Energy Source of the Past and Future*, American Chemical Society, Washington, DC, 1987.
156. E. Fitzer and R. Weiss, in *Processing and Uses of Carbon Fibre Reinforced Plastics*, VDI-Verlag GmbH, Düsseldorf, Germany, 1981, p. 45.
157. W. P. Hoffman, W. C. Hurley, T. W. Owens, and H. T. Phan, *J. Mat. Sci. 26*, 4545 (1991).
158. B. D. Epstein, E. Dalle-Molle, and J. S. Mattson, *Carbon 9*, 609 (1971).
159. (a) K. Kinoshita and J. Bett, *Carbon 11*, 237 (1973); (b) K. Kinoshita and J. A. S. Bett, *Carbon 11*, 403 (1973).
160. N. L. Weinberg and T. B. Reddy, *J. Appl. Electrochem. 3*, 73 (1973).
161. (a) R. E. Panzer and P. J. Elving, *Electrochim. Acta 20*, 635 (1975); (b) R. E. Panzer and P. J. Elving, *J. Electrochem. Soc. 119*, 864 (1972).
162. J. P. Randin and E. Yeager, *Electroanal. Chem. Interf. Electrochem. 58*, 313 (1975).
163. (a) Y. Oren, H. Tobias and A. Soffer, *J. Electroanal. Chem. 162*, 87 (1984); (b) Y. Oren and A. Soffer, *J. Electroanal. Chem. 186*, 63 (1985); (c) M. Yaniv and A. Soffer, *J. Electrochem. Soc. 123*, 506 (1976).
164. V. Markovic, N. Vulevic and D. Markovic, *Fuel 68*, 1039 (1989).
165. R. Bowling, R. T. Packard, and R. L. McCreery, *Langmuir 5*, 683 (1989).
166. C. T. Ho and D. D. L. Chung, *Carbon 28*, 521 (1990).

167. G. Urry, *Elementary Equilibrium Chemistry of Carbon*, Wiley, New York, 1989.
168. G. M. K. Abotsi and A. W. Scaroni, *Carbon 28*, 79 (1990).
169. I. Mochida, M. Ogaki, H. Fujitsu, Y. Komatsubara, and S. Ida, *Fuel 64*, 1054 (1985).
170. Y. Ishikawa, L. G. Austin, D. E. Brown, and P. L. Walker, Jr., in *Chemistry and Physics of Carbon*, Vol. 12 (P. L. Walker, Jr., and P. A. Thrower, eds.), Marcel Dekker, New York, 1975, p. 39.
171. E. Berl, *Trans Faraday Soc. 34*, 1040 (1938).
172. F. J. Long and K. W. Sykes, *J. Chim. Phys. 47*, 361 (1950).
173. R. C. Seymour and J. C. Wood, *Proceedings of 3rd Conference on Industrial Carbon and Graphite*, Society of Chemical Industry, London, 1971, p. 264.
174. V. N. Marinov, *Fuel 56*, 165 (1977).
175. I. M. K. Ismail and P. L. Walker, Jr., *Carbon 27*, 549 (1989).
176. B. C. Gates, J. R. Katzer, and G. C. A. Schuit, *Chemistry of Catalytic Processes*, McGraw-Hill, New York, 1979, p. 383.
177. (a) M. L. Gray, R. W. Lai, and A. W. Wells, *ACS Preprints, Div. Fuel Chem. 36* (2), 804 (1991); (b) J. Attar, U.S. Patent 4,597,769 (1986).
178. S. Mazur, T. Matusinovic, and K. Cammann, *J. Amer. Chem. Soc. 99*, 3888 (1977).
179. J. P. Sibilia, *A Guide to Materials Characterization and Chemical Analysis*, VCH Publishers, New York, 1988, p. 96.
180. (a) L. Eberson, in *The Chemistry of Carboxylic Acids and Esters* (S. Patai, ed.), Wiley Interscience, London, 1969, p. 53; (b) H. Lund, in *The Chemistry of the Hydroxyl Group* (S. Patai, ed.), Wiley Interscience, London, 1971, p. 253.
181. (a) J. M. Campelo, A. Garcia, J. M. Gutierrez, D. Luna, and J. M. Marinas, *Coll. Surf. 8*, 353 (1984); (b) J. M. Campelo, A. Garcia, J. M. Gutierrez, D. Luna, and J. M. Marinas, *J. Coll. Interf. Sci. 102*, 107 (1984); (c) J. M. Campelo, A. Garcia, J. M. Gutierrez, D. Luna, and J. M. Marinas, *J. Coll. Interf. Sci. 95*, 544 (1983); (d) J. M. Campelo, A. Carcia, J. M. Gutierrez, D. Luna, and J. M. Marinas, *Can. J. Chem. 61*, 2567 (1983); (e) J. M. Campelo, A. Garcia, D. Luna, and J. M. Marinas, *Can. J. Chem. 62*, 638 (1984); (f) J. M. Campelo, A. Garcia, D. Luna, and J. M. Marinas, *Afinidad 39*, 325 (1982); (g) J. M. Campelo, A. Garcia, D. Luna, and J. M. Marinas, *Afinidad 39*, 61 (1982).

182. L. Petrakis, P. L. Meyer, and G. L. Jones, *J. Phys. Chem. 84*, 1029 (1980).
183. J. Kijenski and S. Malinowski, *J. Chem. Soc. Faraday Trans. I 74*, 250 (1978).
184. B. D. Flockhart, in *Surface and Defect Properties of Solids*, Vol. 2, Chemical Society, London, 1973, p. 69.
185. A. Ghorbel, C. Hoang-Van, and S. J. Teichner, *J. Catal. 30*, 298 (1973).
186. *Extd. Abstracts Graphite Intercalation Compounds* (M. S. Dresselhaus, G. Dresselhaus, and S. A. Solin, eds.), Materials Research Society, Pittsburgh (1986).
187. J. A. Davis, R. O. James, and J. O. Leckie, *J. Coll. Interf. Sci. 63*, 480 (1978).
188. (a) G. A. Parks, *Chem. Rev. 65*, 177 (1965); (b) G. A. Parks and P. L. de Bruyn, *J. Phys. Chem. 66*, 967 (1962).
189. S. Brunauer, P. H. Emmett, and E. Teller, *J. Amer. Chem. Soc. 60*, 309 (1938).
190. O. Stern, *Z. Elektrochem. 30*, 508 (1924).
191. P. H. Emmett, *Chem. Rev. 43*, 69 (1948).
192. M. M. Dubinin, *Chem. Rev. 60*, 235 (1960).
193. J. T. G. Overbeek, in *Colloid Science*, Vol. 1 (H. R. Kruyt, ed.), Elsevier, New York, 1949, p. 235.
194. C. H. Giles, T. H. MacEwan, S. N. Nakhwa, and D. Smith, *J. Chem. Soc. London 1960*, 3973.
195. J. O'M. Bockris and K. N. Reddy, *Modern Electrochemistry*, Vols. 1 and 2, Plenum Press, New York, 1977.
196. (a) A. M. James, in *Surface and Colloid Science*, Vol. 11 (R. J. Good and R. R. Stromberg, eds.), Plenum Press, New York, 1979, p. 121; (b) R. O. James and G. A. Parks, in *Surface and Colloid Science*, Vol. 12 (E. Matijevic, ed.), Plenum Press, New York, 1982, p. 119.
197. J. J. Lewnard, E. E. Petersen, and C. J. Radke, *J. Chem. Soc. Faraday Trans. I 84*, 3927 (1988).
198. S. J. Gregg and K. S. W. Sing, *Adsorption, Surface Area and Porosity*, 2nd Ed., Academic Press, New York, 1982.
199. H. W. Chen, J. M. White, and J. G. Ekerdt, *J. Catal. 99*, 293 (1986).
200. P. Gallezot, D. Richard, and G. Bergeret, *ACS Symp. Ser. 437*, 150 (1990).

201. J. R. Anderson, *Structure of Metallic Catalysts*, Academic Press, London, 1975.
202. G. S. Rellick, P. A. Thrower, and P. L. Walker, Jr., *Carbon 13*, 71 (1975).
203. J. Lyklema, *J. Electroanal. Chem. 18*, 341 (1968).
204. (a) D. W. van Krevelen and P. J. Hoftyzer, *Properties of Polymers: Correlations with Chemical Structure*, Elsevier, Amsterdam, 1972, p. 211; (b) M. M. Breuer, in *Polymer Science* (A. D. Jenkins, ed.), North-Holland Publ., Amsterdam, 1972, p. 1135; (c) D. A. Seanor, in *Polymer Science* (A. D. Jenkins, ed.), North-Holland Publ., Amsterdam, 1972, p. 1187; (d) P. G. de Gennes, *Adv. Coll. Interf. Sci. 27*, 189 (1987).
205. D. Farin and D. Avnir, in *The Fractal Approach to Heterogeneous Chemistry* (D. Avnir, ed.), Wiley, New York, 1989, p. 271.
206. J. A. Schellman, *J. Phys. Chem. 57*, 472 (1957).
207. J. M. Duxbury, in *Humic Substances II* (M. H. B. Hayes, P. Mac-Carthy, R. L. Malcolm, and R. S. Swift, eds.), Wiley, New York, 1989, p. 593.
208. S. Ergun, J. B. Yasinsky, and J. R. Townsend, *Carbon 5*, 403 (1967).
209. (a) P. N. Cheremisinoff and A. C. Morresi, in *Carbon Adsorption Handbook* (P. N. Cheremisinoff and F. Ellerbusch, eds.), Ann Arbor Science, Ann Arbor, MI, 1978, p. 1; (b) J. T. Cookson, Jr., in *Carbon Adsorption Handbook* (P. N. Cheremisinoff and F. Ellerbusch, eds.), Ann Arbor Science, Ann Arbor, MI, 1978, p. 241; (c) C. P. Huang, in *Carbon Adsorption Handbook* (P. N. Cheremisinoff and F. Ellerbusch, eds.), Ann Arbor Science, Ann Arbor, MI, 1978, p. 281.
210. (a) M. O. Corapcioglu and C. P. Huang, *Carbon 25*, 569 (1987); M. O. Corapcioglu and C. P. Huang, *Wat. Res. 21*, 1031 (1987).
211. (a) D. W. Fuerstenau, *Pure Appl. Chem. 24*, 135 (1970); (b) T. W. Healy, A. P. Herring, and D. W. Fuerstenau, *J. Coll. Interf. Sci. 21*, 435 (1966); (c) T. W. Healy and D. W. Fuerstenau, *J. Coll. Interf. Sci. 20*, 376 (1965); (d) J. A. Yopps and D. W. Fuerstenau, *J. Coll. Interf. Sci. 19*, 61 (1964); (e) A. M. Gaudin and D. W. Fuerstenau, *Trans. AIME 202*, 66 (1955); (f) A. M. Gaudin and D. W. Fuerstenau, *Trans. AIME 202*, 698 (1955).
212. J. P. Brunelle, *Pure Appl. Chem. 50*, 1211 (1978).
213. H. A. Benesi, R. M. Curtis, and H. P. Studer, *J. Catal. 10*, 328 (1968).

214. Y. G. Berube and P. G. de Bruyn, *J. Coll. Interf. Sci. 27*, 305 (1967).

215. (a) S. M. Ahmed, *J. Phys. Chem. 73*, 3546 (1969); (b) P. H. Tewari and A. B. Campbell, *J. Coll. Interf. Sci. 55*, 531 (1976).

216. A. Kitahara and A. Watanabe (eds.), *Electrical Phenomena at Interfaces: Fundamentals, Measurements and Applications*, Marcel Dekker, New York, 1984.

217. D. Fairhurst and V. Ribitsch, *ACS Symp. Ser. 472*, 337 (1991).

218. L. Sawaroski, *Solid–Liquid Separation Processes and Technology*, Plenum Press, New York, 1985, p. 127.

219. (a) R. E. Marganski and R. L. Rowell, *Energy Fuels 2*, 132 (1988); (b) B. J. Marlow and R. L. Rowell, *Energy Fuels 2*, 125 (1988); (c) S. R. Vasconcellos and R. L. Rowell, *Carbon 25*, 97 (1987).

220. (a) J. T. Stock, *ACS Symp. Ser. 390*, 1 (1989); (b) K. J. Laidler, *ACS Symp. Ser. 390*, 63 (1989); (c) B. E. Conway, *ACS Symp. Ser. 390*, 152 (1989).

221. H. P. van Leeuwen and J. Lyklema, in *Modern Aspects of Electrochemistry*, Vol. 17 (J. O'M. Bockris, B. E. Conway, and R. E. White, eds.), Plenum Press, New York, 1986, p. 411.

222. L. Meites and P. Zuman, *Electrochemical Data*, Wiley, New York, 1974.

223. H. Strehlow, in *Investigation of Rates and Mechanisms of Reaction*, 2nd Ed. (S. L. Friess, E. S. Lewis, and A. Weissberger, eds.), Wiley Interscience, New York, 1963, p. 799.

224. S. R. Morrison, *The Chemical Physics of Surfaces*, 2nd Ed., Plenum Press, New York, 1990, p. 297.

225. D. E. Weiss, *Proceedings of 5th Conference on Carbon*, Vol. 1, Pergamon Press, New York, 1962, p. 65.

226. M. Spiro, *J. Chem. Soc. Faraday Trans. I 75*, 1507 (1979).

227. M. Pourbaix, *Atlas of Electrochemical Equilibria*, Pergamon Press, London, 1966.

228. M. K. Karapetyants, *Thermodynamic Constants of Inorganic and Organic Compounds*, Ann Arbor Science, Ann Arbor, MI, 1970.

229. G. A. Somorjai, *Chemistry in Two Dimensions: Surfaces*, Cornell University, Ithaca, NY, 1981.

230. A. T. Hubbard, *Chem. Rev. 88*, 633 (1988).

231. D. Graham, *J. Phys. Chem. 59*, 896 (1955).

232. J. S. Mattson, H. B. Mark, Jr., M. D. Malbin, W. J. Weber, Jr., and J. C. Crittenden, *J. Coll. Interf. Sci. 31*, 116 (1969).

233. V. Y. Glushchenko, T. G. Levagina, and A. A. Pershko, *Koll. Zh.* *37* (1), 134 (1975).
234. O. P. Mahajan, A. Youssef, and P. L. Walker, Jr., *Sep. Sci. Technol.* *13*, 487 (1978).
235. B. R. Puri, in *Activated Carbon Adsorption of Organics from the Aqueous Phase, Vol. 1* (I. H. Suffet and M. J. McGuire, eds.), Ann Arbor Science, Ann Arbor, MI, 1980, p. 353.
236. O. P. Mahajan, C. Moreno-Castilla, and P. L. Walker, Jr., *Sep. Sci. Technol.* *15*, 1733 (1980).
237. T. Asakawa and K. Ogino, *J. Coll. Interf. Sci.* *102*, 348 (1984).
238. Y. Kaneko, M. Abe, and K. Ogino, *Coll. Surf.* *37*, 211 (1989).
239. L. K. Wang, R. P. Leonard, M. H. Wang, and D. W. Goupil, *J. Appl. Chem. Biotechnol.* *25*, 491 (1975).
240. D. O. Cooney and J. Wijaya, in *Fundamentals of Adsorption* (A. I. Liapis, ed.), Proceedings of 2nd Engineering Foundation Conference, United Engineering Trustees, New York, 1987, p. 185.
241. T. Asakawa, K. Ogino, and K. Yamabe, *Bull. Chem. Soc. Jpn.* *58*, 2009 (1985).
242. T. M. Ward and F. W. Getzen, *Env. Sci. Technol.* *4*, 64 (1970).
243. R. W. Coughlin and F. S. Ezra, *Env. Sci. Technol.* *2*, 291 (1968).
244. V. L. Snoeyink, W. J. Weber, Jr., and H. B. Mark, Jr., *Env. Sci. Technol.* *3*, 918 (1969).
245. R. W. Coughlin and R. N. Tan, *Water—1968*, Chem. Eng. Prog. Symp. Ser. No. 90, Vol. 64, 207 (1969).
246. R. W. Coughlin, F. S. Ezra, and R. N. Tan, *J. Coll. Interf. Sci.* *28*, 386 (1968).
247. J. C. Duchet, E. M. van Oers, V. H. J. de Beer, and R. Prins, *J. Catal.* *80*, 386 (1983).
248. F. J. Derbyshire, V. H. J. de Beer, G. M. K. Abotsi, A. W. Scaroni, J. M. Solar, and D. J. Skrovanek, *Appl. Catal.* *27*, 117 (1986).
249. P. Ehrburger and J. Lahaye, *ACS Symp. Ser.* *303*, 310 (1986).
250. D. Richard and P. Gallezot, in *Preparation of Catalysts IV* (B. Delmon, P. Grange, P. A. Jacobs, and G. Poncelet, eds.), Elsevier, Amsterdam, 1987, p. 71.
251. J. P. R. Vissers, S. M. A. M. Bouwens, V. H. J. de Beer, and R. Prins, *Carbon 25*, 485 (1987).
252. A. A. Chen, M. A. Vannice, and J. Phillips, *J. Phys. Chem. 91*, 6257 (1987).

253. A. Guerrero-Ruiz, I. Rodriguez-Ramos, F. Rodriguez-Reinoso, C. Moreno-Castilla, and J. D. Lopez-Gonzalez, *Carbon 26,* 417 (1988).

254. G. M. K. Abotsi and A. W. Scaroni, *Fuel Process. Technol. 22,* 107 (1989).

255. C. Prado-Burguete, A. Linares-Solano, F. Rodriguez-Reinoso, and C. Salinas-Martinez de Lecea, *J. Catal. 115,* 98 (1989).

256. J. M. Solar, F. J. Derbyshire, V. H. J. de Beer, and L. R. Radovic, *J. Catal. 129,* 330 (1991).

257. C. A. Leon y Leon and L. R. Radovic, *Extd. Abstracts 20th Biennial Conference on Carbon,* Santa Barbara, CA, 1991, p. 84.

258. P. L. Walker, Jr., *Proceedings of Fifth London International Carbon and Graphite Conference,* Society of Chemical Industry, London, 1978, p. 427.

259. P. Ehrburger, *Adv. Coll. Interf. Sci. 21,* 275 (1984).

260. H. Jüntgen, *Fuel 65,* 1436 (1986).

261. C. A. Leon y Leon, S. Chakka, and L. R. Radovic, in *Introduction to Carbon Technology* (H. Marsh, ed.), Elsevier, Amsterdam, in preparation (1993).

262. F. Carrasco-Marin, J. M. Solar and L. R. Radovic, *Proceedings of International Carbon Conference,* Paris, France, 1990, p. 672.

263. I. W. Smith, *Fuel 57,* 409 (1978).

264. O. P. Mahajan, R. Yarzab, and P. L. Walker, Jr., *Fuel 57,* 643 (1978).

265. L. R. Radovic, H. Jiang, and A. A. Lizzio, *Energy Fuels 5,* 68 (1991).

266. Z.-B. Zhu, T. Furusawa, T. Adschiri, and T. Nozaki, *ACS Preprints, Div. Fuel Chem. 34* (1), 87 (1989).

267. L. R. Radovic, A. A. Lizzio, and H. Jiang, in *Fundamental Issues in Control of Carbon Gasification Reactivity* (J. Lahaye and P. Ehrburger, eds.), Kluwer, Dordrecht, The Netherlands, 1991, p. 235.

268. F. Kapteijn, R. Meijer, B. van Eck, and J. A. Moulijn, in *Fundamental Issues in Control of Carbon Gasification Reactivity* (J. Lahaye and P. Ehrburger, eds.), Kluwer, Dordrecht, The Netherlands, 1991, p. 221.

269. J. M. Calo and P. J. Hall, in *Fundamental Issues in Control of Carbon Gasification Reactivity* (J. Lahaye and P. Ehrburger, eds.), Kluwer, Dordrecht, The Netherlands, 1991, p. 329.

270. K. J. Hüttinger and J. S. Nill, *Carbon 28*, 457 (1990).
271. J. Wang and B. McEnaney, *Extd. Abstracts 19th Biennial Conference on Carbon*, University Park, PA, 1989, p. 590.
272. K. H. van Heek, *Proceedings of International Carbon Conference*, Newcastle upon Tyne, UK, 1988, p. 252.
273. F. Kapteijn, R. Meijer, and J. A. Moulijn, *ACS Preprints, Div. Fuel Chem. 36* (3), 906 (1991).
274. M. H. Back, *Carbon 29*, 1290 (1991).
275. C. A. Coulson, *Proceedings of 4th Conference on Carbon*, Pergamon Press, New York, 1960, p. 215.
276. P. L. Walker, Jr., private communication.
277. J. L. Johnson, in *Coal Gasification*, Adv. Chem. Ser. No. 131, American Chemical Society, Washington, DC, 1974, p. 145.
278. H. Yamashita, A. Tomita, H. Yamada, T. Kyotani, and L. R. Radovic, *Energy Fuels 7*, 85 (1993).
279. H. Yamashita, H. Yamada, and A. Tomita, *Appl. Catal. 78*, L1 (1991).

Index